T0214667

BEYOND FREE MARKET

This book explores the causes and consequences of market failure in bridging societal differences to create a shared economy. It questions the current world order and evaluates socio-economic gains in reference to the social origins of the economic agents.

With a need to counterbalance economic growth with social equality and environmental sustainability, the book proposes innovative approaches to address key questions on the contemporary global economy such as, "Is the Global socio-economic order supportive of the pursuit of rational and enlightened self-interest?", "Is it a unipolar power centre and neoliberal economic policy regime?", "Can the system reinvent itself?", etc. One approach encourages going back to the golden past and making things "great again", insisting that history has ended, and the failures of old global institutions be blamed on the "Clash of Civilizations". Another approach advocates giving up the intellectual comfort zone of elegant but irrelevant neo-liberal explanations of global challenges and asking new questions that take academic debate to the public square. The book examines the internal challenges and contradictions that cause disintegration and proposes alternative ideas and practices in moving the global community beyond the free market regime.

The book will appeal to students and academics of development studies, political economy, political science, sociology, as well as policymakers and public opinion makers interested in creating a new egalitarian global society.

Fayyaz Baqir is a visiting scholar at the University of Ottawa. He served as senior advisor on civil society at the United Nations, and CEO of Trust for Voluntary Organizations. He received top contributors' awards from UNDP's global poverty reduction network.

Sanni Yaya is Full Professor of Economics and Global Health in the School of International Development and Global Studies at the University of Ottawa and holds the Senghor Research Chair in Health and Development.

The Dynamics of Economic Space

This series aims to play a leading international role in the development, promulgation and dissemination of new ideas in economic geography. It has as its goal the development of a strong analytical perspective on the processes, problems and policies associated with the dynamics of local and regional economies as they are incorporated into the globalizing world economy. In recognition of the increasing complexity of the world economy, the Commission's interests include: industrial production; business, professional and financial services, and the broader service economy including e-business; corporations, corporate power, enterprise and entrepreneurship; the changing world of work and intensifying economic interconnectedness.

BEYOND FREE MARKET

Social Inclusion and Globalization

Edited by
Fayyaz Baqir and
Sanni Yaya

Routledge
Taylor & Francis Group

LONDON AND NEW YORK

First published 2021
by Routledge
2 Park Square, Milton Park, Abingdon, Oxon OX14 4RN

and by Routledge
605 Third Avenue, New York, NY 10158

Routledge is an imprint of the Taylor & Francis Group, an informa business

© 2021 selection and editorial matter, Fayyaz Baqir and Sanni Yaya; individual chapters, the contributors

British Library Cataloguing-in-Publication Data
A catalogue record for this book is available from the British Library

Library of Congress Cataloging-in-Publication Data
A catalog record has been requested for this book

ISBN: 978-0-367-55328-9 (hbk)
ISBN: 978-1-003-09302-2 (ebk)
ISBN: 978-0-367-55333-3 (pbk)

Typeset in Bembo
by KnowledgeWorks Global Ltd.

Dedicated to time, the greatest teacher, that invites us to discover truth veiled in mystery.

CONTENTS

LIST OF FIGURES

LIST OF TABLES

LIST OF CONTRIBUTORS

Furqan Asif is an environmental social scientist with grounded research experience in coastal, small-scale fisheries and communities (Cambodia), marine protected areas (Philippines), human migration, social wellbeing, and social resilience. Currently, he is a postdoctoral researcher within the Environment Policy Group at Wageningen University & Research (Netherlands) focusing on building upon the theory and praxis of aquaculture governance via the global implementation of an aquaculture governance indicator framework. He holds a PhD in International Development (University of Ottawa), a Master of Environmental Science (University of Toronto), and an Honours Bachelor of Science in Biology/Environmental Science (Western University).

Juan Velásquez Atehortúa is a PhD in Human Geography and Associate Professor in Gender Studies. His areas of research focus on participatory democracy from the perspective of *barrio* women and other subaltern constellations mostly in Latina America and the Caribbean. He has published his works in the tradition of participatory action research combining scholarly works in academic journals in English, Spanish, and Swedish, and building an academic video-archive with ethnographies from subaltern group actions and internal debates collected in support of these movements in Venezuela, Colombia, Bolivia, and Nicaragua in the Americas and Sweden in Scandinavia.

Fayyaz Baqir is a visiting Professor at the School of International Development and Global Studies (SIDGS) University of Ottawa. He served as an O'Brien Fellow at McGill University, and as visiting scholar at the Department of Cultural Sciences, Gothenburg University (Sweden), and the School of European and International Public Law at Tilburg University (the Netherlands) in the past. His professional and academic interests include participatory development, human rights, aid effectiveness, poverty alleviation, and social accountability. He co-designed and taught cross-border video conference-based courses on themes relating to Justice and Peace, Social Change and Politics of Human Development in Pakistan for Georgetown University, Harvard University, Wellesley College, and Fatima Jinnah University. Earlier, he served as Senior Advisor on Civil Society for the United Nations in Pakistan. He received Top Contributors Award from the UNDP's Global Poverty Reduction Network in 2007 and 2008.

He has authored numerous journal articles, conference papers, book chapters, training manuals, and books on participatory development, inclusive govern-ance, and poverty alleviation. His recent publications include Better Spending for Localizing Global Sustainable Development Goals: Examples From the Field, Routledge London, November 2019, and Poverty Alleviation and Poverty of Aid-Pakistan, Routledge London, August 2018.

Tricia Glazebrook is a Professor of Philosophy in the School of Politics, Philosophy, and Public Affairs at Washington State University. She is a Heidegger scholar publishing on his critique of science and technology, and implications of his work for environment and climate issues, gender and economics. Broader areas of expertise include ecofeminism, philosophy of international development, and military ethics. Long-standing research on women's agriculture and food security in Ghana currently focuses on climate finance and policy. She also studies oil-driven conflict in sub-Saharan Africa and tension between oil development and food security as Africa works to meet Sustainability Goals on poverty and hunger. She is Deputy Chair of Gender CC: Women for Climate Justice, an international non-profit with active projects in the global South to integrate women into urban climate policy.

Josh Hawley is a working-class organizer from Ottawa and a founding member of the Herongate Tenant Coalition, an independent organization fighting displacement and building class power. Josh is currently pursuing a PhD in sociology at Carleton University. His MA in Cultural Studies (Queen's University) took a critical look at co-operative housing as a neoliberal tool of tenant self-management. His archival research of the Montreal neighbourhood of Milton-Parc resulted in the book Villages in Cities: Community Land Ownership, Cooperative Housing, and the Milton Parc Story (co-edited with Dimitri Roussopoulos, Black Rose Books, 2019), and the exhibit Milton-Parc: How We Did It at the Canadian Centre for Architecture in Montreal. Josh is helping to build multiple legal cases against a multi-billion dollar corporate landlord as well as the City of Ottawa.

Nima Hussein is an Ottawa-based organizer and a founding member of the Herongate Tenant Coalition. Her degree from the University of Ottawa explores global development theory, and she is a member of the BlackThots Collective, a reading circle dedicated to dissecting Black feminist thought. Nima is a fiction contributor at The Drinking Gourd, an online literary magazine. She is currently involved in organizing with her neighbours against the largest mass displacement of urban Black and Brown bodies in Canadian history in her home neighbourhood of Herongate. As a result of the 2018 phase of these evictions, Nima is working to build a groundbreaking collective rights case that challenges the western, individualized conceptualization of human rights in capitalist states.

Carolyn Laude has worked at the First Nation, post-secondary education, and senior government levels on a broad range of Indigenous affairs policy files. She is presently completing a PhD in Legal Studies at Carleton University. Carolyn's research projects explore the tension between the Indigenous and settler-colonial life worlds, which hold distinct understandings of rights, land, and self-determination. She recently examined the Tsilhqot'in Nation decision to explore how settler-colonial modes of reconciliation account for and/or contradict Indigenous inherent rights and socio-legal ordering. Carolyn's research interests focus on decolonizing and reconceptualizing dominant modes of reconciliation in support of Indigenous and settler-colonial right relations and amicable co-existence. She holds a Master's in Legal Studies and a Bachelor of Arts Honours in Political Science.

Abdullah Al Mamun is a highly skilled evaluation, research, and data science professional with over ten years' experience in several global institutions Afghanistan, Bangladesh, India, Nepal, Tajikistan, UK, and USA. He holds an MA in Sustainable International Development from the Brandeis University, United States. He also obtained MSc and BSc in Statistics from the Jagannath University College, Bangladesh. His research interest is in poverty measurement, economic growth and development, and public health. He has strong interest in developing effective and efficient policies based on scientific evidence. He has co-authored publications in these research areas and is currently a PhD candidate at the University of Ottawa.

Adrian Murray is a post-doctoral fellow in the Department of Integrated Studies in Education at McGill University. His research explores processes of social movement learning and knowledge production around struggles for access to water services in the face of ongoing austerity across the global South and North. In this and other work Adrian is affiliated with the Municipal Services Project, a research project that explores alternatives to the privatization of public services in Africa, Asia, and Latin America and the Council of Canadians' Blue Planet Project. He is also an active trade unionist.

Syed Mansoob Murshed is a Professor of the Economics of Peace and Conflict at the International Institute of Social Studies (ISS), Erasmus University in the Netherlands and Professor of Economics at Coventry University, UK. He was the first holder of the rotating Prince Claus Chair in Development and Equity in 2003. He was a Research Fellow at UNU/WIDER Helsinki where he ran

Projects on Globalization and Vulnerable Economies and Why Some Countries Avoid Conflict, While Others Fail. He also ran a project on The Two Economies of Ireland, financed by the International Fund for Ireland at the Northern Ireland Economic Research Centre (NIERC), Belfast. He is the author of nine books and over 160 refereed journal papers and book chapters. His latest monograph is on the Resource Curse (2018). He is on the Editorial boards of the Journal of Peace Research, Peace Economics, Peace Science and Public Policy, and Civil Wars. His research interests are in the economics of conflict, resource abundance, aid conditionality, political economy, macroeconomics, and international economics.

Syed Sajjadur Rahman is a Senior Fellow and visiting Professor at the School of International Development and Global Studies at the University of Ottawa. During 1992–2013. He worked in the Canadian International Development Agency (CIDA) (now part of Global Affairs Canada), Government of Canada, in senior positions including as Associate Vice-President, Policy, Regional Director-General, Asia, and the Executive Director of the Development Finance Innovation Working Group. Dr. Rahman has a PhD in Economics from Carleton University, Ottawa, Canada. His research interest includes development partnerships with middle-income economies.

Rasul Bakhsh Rais is a Pakistani political scientist and Professor of Political Science at the Lahore University of Management Sciences (LUMS). He has several publications to his credit, and is currently researching Western Pakistan's geopolitical situation, Pakistan's political dynamics, security, and nuclear weapons political issues. Rasul has been affiliated with the Department of International Relations at Quaid-i-Azam University. Rasul also taught Pakistan Studies at Columbia University in 1991–94. He is currently teaching political science courses at the Lahore University of Management Sciences (LUMS) and instructing courses on the theories of democratic transition, American government and politics, and comparative politics.

Susan Spronk is an Associate Professor in the School of International Development and Global Studies at the University of Ottawa. Her research focuses on the experience of development in Latin America, more specifically the impact of neoliberalism on the transformation of the state and the rise of anti-privatization movements in the Andean region. She has published various articles and chapters on social movements, working-class formation, as well as water and sanitation politics in Latin America. She is also a research associate with the Municipal Services Project, has been a community organizer for over 20 years, and is an active trade unionist.

Kazue Takamura is a Faculty Lecturer at the Institute for the Study of International Development, McGill University. Her research primarily focuses on Asia's emerging migrant surveillance regimes, especially the overlooked intersection between neoliberal labour markets and illiberal migrant punishment including immigration detention. Her works also pay attention to the gendered patterns of labour mobility and immobility that are produced by the neoliberal migrant surveillance regimes.

Ogochukwu Udenigwe is a PhD candidate and SSHRC scholar in International Development and Global Studies at the University of Ottawa. Her research explores strategies to improve maternal and child health in the Global South. Ogochukwu has co-authored 15 papers published in peer-reviewed journals and books. She earned her MSc degree in Family Relations and Human Development from the University of Guelph and two Bachelor's degrees in Family Social Sciences and Chemistry/Zoology from the University of Manitoba.

Sanni Yaya is Full Professor of Economics and Global Health, Director and Associate Dean of the School of International Development and Global Studies. His work focuses on a broad array of multidisciplinary topics in development and global health. This includes cross-cutting research and publications in disciplines of public-private partnerships, global governance, economic development, and global health. He has a strong interest in large-scale evidence (in particular randomized trials) and analyses of large survey data sets.

Helena Yeboah is a PhD candidate at the University of Ottawa pursuing the International Development and Global Studies program. She holds dual Master's in Applied Economics and International Development Studies from the Ohio University, United States and obtained her BA in Social Sciences (Economics major) degree at the University of Cape Coast, Ghana. Her research interest is in poverty reduction, economic growth and development, and maternal and child health. She has co-authored publications in these research areas and is currently a Teaching Assistant at the University of Ottawa.

1

INTRODUCTION

Fayyaz Baqir

Freedom, science, and the wealth of nations

The essays presented in this book agree on one key premise, i.e. there are many worlds in the "One World" that we are in the habit of seeing. There is no singularity of truth, and truth happens to be pluralist in nature. We cannot wipe off the logic of nature with the logic of power and power cannot subdue the humility that creates awareness of "knowing what one does not know."

The Age of Enlightenment ushered in an era of unprecedented human development based on reason, science, and individual freedom. This search for human freedom led to the creation of the free market, laying the foundation of the industrial revolution and the institutionalization of elected forms of government. However, the walk to freedom took place on very slippery ground. The foundations of the free market were invented during the colonial era. Indeed, pursuit of free markets was accompanied by the pursuit of colonies, the slave trade, violent protest movements of the working class for social justice, and global wars. How could human beings, led by the impeccable and extremely trustworthy faculty of reason, give birth to consequences that could possibly be done in the state of extreme ignorance? We need to ask why the markets and democracy that enable us to increase the range of our choices and make the most rational decisions, and which guarantee the greatest happiness for the greatest number, end up creating maximum suffering for the maximum number (Sen 2000). It seems we need to start searching for dark forces at work inside an individual or a society in the age of freedom. We needed to know the limitations in the way of our reasoning.

The limitations of our reasoning originated in the nature of scientific knowledge that we employed. Science provided us precision at the expense of wholeness. It changed the relationship between the body, mind, and soul. The capitalist outlook celebrated the body, suspected the soul, and commodified the mind; relegating consciousness to the pursuit of utilitarian ends through the narrowly carved path of "positive" social science. The capitalist mind saw the planet and the universe as its mirror image, a body without a soul. Celebration of bodies triggered off a never-ending quest for colonization, possession, control, and consumption of land and natural habitat of pre-capitalist communities, kingdoms, and empires all over the world. In the course of its rise and expansion during the next four centuries, the global capitalist empire countered moments of strife, conflict, chaos, and disillusionment, and finally it is coming closer to the moment of self-annihilation. The capitalist order has been dangerously disruptive and it expanded by demolishing the organic relationships between human communities, and the holistic relationship between humans and nature, and body and soul. It converted every manifestation of nature and every dimension of human activity into a commodity, into an object for drawing utilitarian pleasures. The underlying contradictions of this reductionist and utilitarian approach have brought us and our natural habitat to the brink of destruction.

Reductionist scientific framework fed into the basic premises of our thought framework and our understanding of freedom of choice, freedom of inquiry and expression, rationality, the free market, and democratic forms of government. The rationalist toolkit included Jeremy Bentham's utilitarianism emphasising individual rationality (Bentham 1907), Adam Smith's (2018) virtuous account of capitalism pointing to the miraculous quality of an invisible hand in allocation and distribution of resources, and a free market dispensing greatest happiness to the greatest number. Optimal choice made by rational consumers was based on the premise that every new unit of consumption creates diminishing marginal utility, so rational consumption will discourage greed and gluttony. Only in the case of money this law did not apply. This was the only exception which undid the balancing role of the market in pre-empting the urge for the concentration of wealth in a few hands and proved true Gandhi's view that the world has enough resources to meet the needs of all people but not enough to satisfy the greed of one individual (Balch 2013).

If we look at the entire modernist discourse, we see exceptions at work in every nook and corner. All narrowly defined scientific truths need to embrace exceptions to gain general validity. This exceptionalism has produced unintended consequences that far exceed the intended consequence. It is time we look at these exceptions and try to find alternatives that pre-empt the negative impacts of these consequences. This calls forth the need to move from a singular concept of truth to a pluralist concept of social reality.

Critique of rational decision making

Conventional economic thought makes the case for its rationale by introducing a distinction between the "positive" and "normative" sciences. Whereas positive science looks at the things as they are, normative science proposes how the things should be based on some normative judgment. The positive view therefore discards "normative" views as wishful thinking, not in line with the ground reality. Let us begin with the positive scientific method. One of the limitations of the positive method is that it offers precision at the cost of holistic understanding of the world. It looks at social reality not as something interpersonal, but private, devoid of any social context and not built on social interaction. Disregarding this social dimension and inherent conflict in social transactions made in the marketplace results in alienation, displaced aggression, violence, and wars (McKenzie 1981). Positive rationality was also used to promote a utilitarian concept of nature which has led us to the brink of extinction due to climate change. The idea of "rational individual choice" has been smuggled into the realm of science due to its strong appeal as simple common sense. However, common sense has both its strengths and limitations. Common science premises were contested in a masterly way by Socrates who questioned ascribing universal applicability to phenomena that have limited validity. He challenged each common sense-based proposition by pointing out the exceptions to the rule and seeking explanation of the exception through his dialogical discourse (Plato and Greene 1927).

Common sense perceptions have been challenged in a profound way by two contemporary thinkers, Nassim Nicholas Taleb in the *Black Swan* and Daniel Kahneman in the book *Thinking Slow, Thinking Fast*. Kahneman pointed out that the human mind works on two tracks, fast and slow. The fast track looks at the similarities and ignores the differences between the phenomenon that it observes, and this furnishes the basis for stereotyping. Thinking fast can both illuminate and obscure the reality that we are trying to understand. According to Kahneman, the limitation of stereotypical thinking can be overcome by looking at the numbers instead of looking at the similarities and differences (Kahneman 2011). But numbers alone may not suffice. Numbers compel us to look at the facts. But as pointed out by Hegel, the fact is identity of something with our ideas whereas truth is the identity of something with itself (Hegel, Baillie and Lichtheim 1967). Taleb has vehemently attacked "professional" views on slavish conformity to stereotyping and lack of attention to unique and least visible possibilities – known as error term in statistics – and named "Black Swan" by him (Taleb 2010). It is important to mention here that the fundamental task of a scientist is to tear the veil of appearance and unveil the mystery of the world around us and continuously challenge our "scientific views" which have taken the forms of ideologies and beliefs.

Another limitation of positive economics is the tautology of reasoning. A case in point is the functioning of the so called invisible hand so persuasively argued by Adam Smith. The invisible hand leads to decision making through the price mechanism. Price is considered the best indicator because it reflects the scarcity of a good or service in the marketplace. So, scarcity determines the price of a good. How do we find the scarcity of a good? Price determines the scarcity. So, we are moving in a circle, but the force of habit provides a big service by making us feel comfortable about this paradox. The concept of scarcity plays a vital role in explaining individual choices made as consumers and producers. The solution to the economic problem, as defined by conventional economics, is making an optimal choice to satisfy unlimited wants through limited means. Price is considered the best guide in evaluating individual choices with reference to the utility yielded by different baskets of goods and services compared to a certain amount of money. However, this method sweeps under the carpet the distinction between the natural and social scarcity of a good or service (Linder and Sensat 1977). Free market fails in making this distinction. Consequently, the free market fails in guaranteeing full employment, fair wages, and mitigation of climate change. It is only through state intervention that these failures can be corrected. However, during past 40 years of neoliberal policies, state has also failed to play a meaningful role. This compels us to revisit the conventional wisdom and go beyond "Free Market" ideology.

In the past, the free market could not correct its own errors to promote sustained and equitable economic growth. The breakdown of the free market resulted in the Great Depression in 1929. Under the New Deal, state expenditure was accepted as a new norm to stabilize economies around full employment. Subsequently, the market's inability to reduce global inequality was recognized and development assistance was introduced as an effort to create a global welfare system under the auspices of the United Nations and Bretton Woods institutions. Two World Wars demonstrated that political decisions of democratic societies are not led by rationality, but that rationality is led by the desire to accumulate; and scientific truth based on reason is led by personal truth based on emotions. This conflict between desire and rationality furnishes the fundamental source of instability from the micro to macro level in the contemporary global community. Half a century after the New Deal, predatory capital staged a comeback in the form of supply side economics. This recipe could not prevent the recession of 2008 and later on led to promises by populist politicians to withdraw from open trade in the global market. Ironically, while the scientific thought deliberated on what determines the value of a good in the society, it did not come up with a conclusive response to the source of this value; there was no conclusive response to what constituted the source of value? Is it nature, labour, or capital? The answer perhaps lies in intellectual humility.

Prices also send another confusing signal to the consumer. Every sale price at a purchase location includes sale tax, indicating the amount that the consumer is paying to the government and is seen by the Libertarians as an infringement on the benefit received by the consumer. But ironically the price does not show the subsidy received by a business due to use of patents created as part of military research or public sector funding of scientific research or subsidies received by a corporation due to government bailout. Tax deduction at the sale point sends a message hundreds of millions of times to all the consumers that they are subsidizing the workers on welfare or claiming health insurance but says nothing about the percentage going to weapon production, war, military research, and corporate support. That is why, as proposed by Kahneman, we need to look at the numbers not images if we want to go beneath the appearance.

"Free Market" ideology also justifies market price as a "natural price" because it is determined on the basis of supply and demand – expressing the free and voluntary choices of consumers and producers in the market. In the first place this natural price reflecting actual scarcities does not exist (Linder and Sensat 1977). Secondly, the assumption that natural price, also known as equilibrium price, clears the market making both the consumers and producers happy by providing them maximum benefit from a transaction does not hold. As shown by Ha Joon Chang (2012), some prices do not settle at equilibrium level even over the course of a century. So, this equilibrium might exist in the imagination of economists or gullible citizens but not in reality. Then same happens in the case of wages.

The myth that the free functioning of the market leads to full employment and fair wages, and the greatest happiness for greatest number came under question in the second quarter of the twentieth century (Bauzon 2016). Stagnation kept haunting the global capitalist economy until the Great Depression of 1929 ushered in the Keynesian solution of public sector spending under a mixed economy labelled as the New Deal (Hanauer 2019). Joseph Schumpeter (2008) pointed out that one way of dealing with stagnation in the market economy was creative destruction. In the post WWII period, American Empire found that war was one way of handling stagnation and cyclical developments. War is a profitable endeavour if conflict and war could be sold to the public mind in the image of peace. Playing on the subconscious uncertainty and fear of citizens in the capitalist world, war was promoted as harbinger of peace. Mass media was handy in cultivating this fear and sustaining arms production and the war machine (Moyn and Wertheim 2019).

As aptly pointed out by Skidelsky (2019), economics can only solve the problems that it imagined having existed a long time ago. It disregards any changes in the social reality. Conventional economics believed that high wages and smooth growth could be ensured if government only kept inflation in check by controlling the money supply. However, printing money does not cause inflation and high levels of employment do not increase wages in contemporary economies. Yet the language of public debate and the wisdom conveyed in economic textbooks remain almost entirely unchanged. The basic psychological assumptions of mainstream (neoclassical) economics including the proposition that flexible wages would ensure full employment has been proved wrong by psychologists and the major economic events of our time. However, the false assumptions of economics have colonized the rest of the academy and have had a profound impact on popular understandings of the world (Skidelsky 2019). As noted by Graeber (2019), economic theory cannot explain the key economic changes in the contemporary global economy and resembles a shed full of broken tools. The problem of how to determine the optimal distribution of work and resources to create high levels of economic growth is simply not the same problem we are now facing, i.e., how to deal with increasing technological productivity, decreasing real demand for labour, and the effective management of work, without also destroying the earth. This demands a different science.

End of the Cold War, neoliberalism, and globalization

The post WWII period witnessed the emergence of a Socialist Bloc, the so called Second World, and free former colonies that together constituted the Third World along with the former colonial powers or the "Free World" known as the First World. The end of WWII gave birth to a new struggle for supremacy between the "Free World" and the Socialist Bloc led by the USA and the USSR, respectively. This conflict was named the "Cold War" and both the USA and USSR tried to win their allies in the Third World, engaging in low-intensity local conflicts, coup d'états, and localized wars through their proxies. This conflict ended with the collapse of Soviet Union in 1992 following the fall of Berlin Wall in 1989. The end of the Cold War did not result in the peace dividend that was expected by the global community. During the post WWII years, an explosion of knowledge took place as the global economy rapidly moved from the first industrial revolution to the fourth industrial revolution. Beginning in the 1980s, the gains made under the New Deal were reversed under neoliberalism (Mann 2020; Bockman 2013). The effectiveness of state for promoting common welfare compared to the "free market" was questioned (Brosio 2008). There was continuation of perpetual tension between the state and the market whereby the business class desired the engagement of the state to protect its interests and uphold freedom of the market from the state when its own interests were challenged (Frieden 2012).

Neoliberalism closely resembles Gore Vidal's eloquent statement on the American system that it is more like "socialism" for the rich and "free enterprise" for the poor (Vidal 2005). The system which was governed by the Bretton Woods institutions got fractured when the value of the dollar was delinked from gold in 1971 due to increased international financial flows and speculative pressure on dollar and other currencies. This led to introduction of flexible exchange rates instead of fixed exchange rates. Increased flows of international finance and a burst of borrowing by the least developed countries (LDCs) led to a debt crisis of unprecedented proportions. The economic shock caused due to the oil price increase by the Organization of Petroleum Exporting Countries (OPEC) led to a simultaneous high rate of inflation and unemployment known as stagflation. During these crises, advanced capitalist economies returned to greater market rule, deregulation, privatization of public enterprises, and tax reforms although the United States followed the "anomaly" of deficit financing to deal with the crisis during this period (Christianto 2018).

The debt and oil crises and the shortcomings of import substitution strategies in less developed economies led to the greater openness of the world economy. The collapse of centrally planned economies removed the remaining barriers standing in the way of the globalization of the world economy. This was followed by the emergence of regional economic blocks. Capitalism now became global and the globe became capitalist. However, neoliberalism has eroded states' authority and ability to regulate their own affairs, while globalization has not ended capitalism's existential problem of recurrent periodic crises hence the need for recurrent state intervention. Neoliberalism, according to some thinkers, is less about stateless markets and more about a market world protected from demands of mass justice and redistributive equality (Iber 2018). Recurring financial and currency crises are the price of open financial markets.

Capitalism has been in perpetual motion from its beginning. While free markets and democracy have held centre stage, the structure, practices, policies, and ideological representation of capitalism have kept changing. Contrary to market orthodoxy, the capitalist system has to be managed both by the invisible hand of the market and the visible hand of the state (Scott 2006; Chavance 2003; Kocka 2010). The questions now staring in the face of global capitalism are about making a choice between a predatory or a pluralist economy, and dealing with the issues of deindustrialization, degrowth, and re-appropriation of the commons as an alternative to glaring inequalities and climate change (Frieden 2012). The neoliberal order has been a political success and an economic failure (Kuttner 2019, Rodrigues & Kuttner, 2019). Contrary to the promises made by the political class that lower tax rates and financialization guaranteed under the neoliberal order would lead to high rates of growth and high standard of living, advanced capitalist economies suffered from deindustrialization, polarization, and a shrinking middle class (Stiglitz 2019).

Does necessity reveal itself as a chance happening?

Economic and social crises in the capitalist system have been caused both by internal factors, failure of the market, and external shocks arising from sudden shortages and price changes, conflicts, wars, and epidemics. Two such events have deepened the crises already underway in the global economy and hastened the process of social struggle to bring about major shifts in the system. These two events are the spread of the COVID-19 pandemic and the murder of George Floyd – a black American at the hands of white police officers. Both these events have highlighted numerous things which defy the logic of private wealth accumulation at the cost of social and economic survival. One thing these events have conclusively demonstrated is the interdependence between the global community – workers, people of diverse identities, businessmen, and politicians – and interdependence between humans and nature. The coronavirus pandemic has given birth to tectonic movement in the global economic order.

The coronavirus pandemic reminisces the words of former secretary of the League of Nations Sir Sultan Mohammad Shah Aga Khan who likened a crisis like this to a World War III. Our enemy this time is our own denial of our interdependence, our own disregard for the truth that as a species and as a part of nature we rise and fall together. This war is extremely difficult to handle because the enemy has taken refuge in our most favourite quarters, our own hearts, minds, and bodies. This pandemic has shown us as a species how much we are dependent on other human beings and on nature. The coronavirus does not recognize the barriers we humans have created between ourselves, our natural habitat, and our intellectual and social heritage that evolved over thousands of years of hard work and rational thinking, for our progress and well-being. We are in trouble because we have mocked our knowledge heritage considering it a treasure trove of ignorance. We consider the difference in race, colour, class, ethnicity, gender, and caste not as a source of diversity and richness, but as an opportunity to pass judgment on the "other" as inferior or superior, good or bad, worthy of compassion or rejection, and to take advantage of their vulnerability. The coronavirus does not recognize any such judgment; it attacks all of us indiscriminately. While we divide ourselves between successful and failed, developed and underdeveloped, hardworking and lazy, superior and inferior, enlightened and obscurantist, high caste and untouchables, secular and religious, scientific minded and superstitious, the coronavirus does not respect such fine differences. It sweeps all of us with one brush. It tells us that we rise and fall together, that our fate and the fate of our planet are intertwined. The coronavirus tells us that we can find safety, security, salvation, freedom, joy, and happiness through others.

Coronavirus is also demolishing the walls between the rich and the poor by pulling down entire economies. It is also savagely suffocating the Free World by making governments in Free Societies borrow a page from totalitarian governments. It has unleashed a conspiracy on us to bring us down to the level of "uncivilized" Indigenous communities. If the crisis persists, our diet habits might shift from supermarket consumption to pure, pristine, and unfree natural food. And the bigger danger is that if this pandemic persists much longer, it might even erase from our memory the accomplishments of the superb consumerist culture of our "free civilization:" innumerable flavours of artificially sweetened yogurt, varieties of broiler chicken, fizzy drinks, fast food, and much more. What will happen to the great and glamorous heritage that we created at the "end of history" and which we were trying to save by going back in time to the utopia of our golden age, where "I" came first and everything else did not matter?

Corona seems to be getting to our jugulars. But we are still not sure who will win in the end: "knowledge of ignorance or ignorance of knowledge?" Perhaps it will help to see how Impala, not a "rational" creature like humans, strive for survival. It is interesting to note that Impala become individualistic during periods of bounty and collectivist during periods of drought. They know the difference between adversity and plenty. On the other hand, humans see horizontal social reality as vertical social reality, social scarcity as natural scarcity, poison as profit, and destruction of our natural habitat as power over nature and development of our being. We buy goods on credit to keep them in the rented space, accumulate wealth to delegate or defer consumption; we reject religion as the opium of the people but accept commodities as the opium of the people (Ali 2008).

The death of George Floyd on the heels of coronavirus pandemic has split open various fissures in the global village. These fissures are caused by the insensitivity of the system to the oppression committed against the global precariat on the basis of differences in race, gender, ethnicity, class, caste, and geographic origin. These prejudices have a long history but so called democratic societies were not addressing these biases due to the corruption of the political class during the age of neoliberalism. The neoliberal order is characterized by an unprecedented free flow of capital across borders, weak nation-states (in both the developing and developed world), and deep divisions among the global working class and between various identity groups. All these things are closely linked. The free flow of capital facilitated and corrupted the neoliberal order (Sachs 2011). It led to a massive outflow of corporate capital from North to South. Due to competition between states and the availability of tax havens, it enabled corporate capitalists to enjoy tax breaks, become super-rich, bribe the political class, hurt the working class in the North and start regime change and support terrorist campaigns in the South, leading to South-North migration. Flight of capital and immigration both benefited the super-rich, but they have successfully projected the image of Northern working class suffering at the hands of Southern workers and immigrants. In my view, this perception did not lead to disillusionment with the neoliberalism but with the impulse of globalization. Expression of shock, anger, and solidarity at George Floyd's death reversed that divisive discourse.

The wave of protests in the wake of Floyd's death has started healing the wounds caused by divisions along identity lines. It has created both opportunity and serious risk. The current crisis is different from all previous crises in clearly demonstrating that freedom, prosperity, and survival are not possible at the individual or identity level. This creates the opportunity for the global precariat to challenge the basic premises of capitalist ideology – the so-called elusive and fabricated concept of "individual" freedom. Ending capitalism means coming out of the cocoon of self- or identity groups. It means prosperity, peace, and freedom are only possible through others and with others. In concrete terms, it means realizing that freedom is only possible together; solidarity is important, but solidarity at the local or identity level means reducing the pain and solidarity at the global level means eliminating the cause of pain and creating a global solidarity economy. A solidarity economy is our only hope and it means re-allocating resources from private to public spending, from arms manufacturing to healthcare and education, from unemployment benefits to universal basic income, from discrimination in wages to equal pay for equal work, from conquering nature to reintegrating with nature, and much more. The risk arises from our habits of thought, the habit of seeing uniformity as the only form of unity and not seeing diversity and pluralism as forms of unity.

Pluralism means that there is something that each identity group needs to get rid of to end capitalism. So, we need to dig deeper. It might mean that capitalism cannot be confined to relationships based on property ownership only. Identity itself can assume the form of capital if it is used to extract an unequal exchange from people of other identity. Ending capitalism in this context would mean ending the practice of taking advantage of the vulnerability of other individuals or groups. Another question to ask is if it can happen voluntarily (without the use of state power and therefore without aspiring to capture power)? On the other hand, is it possible to make it happen by capturing power? Did State power-based socialism succeed in that? It takes us to a deeper level of questioning. Would transformation without taking power mean that we agree on a "process" to reform and restrain capitalism without aspiring to end the nation state? The spread of coronavirus and the death of George Floyd have created two powerful sparks. It is not certain if these sparks will create a prairie fire but if it happens, it will hasten the process of social transformation.

The way forward is to speak of a pluralist social discourse that accepts room for the State, the Market, and the Commons in the economic and social sphere(s). This pluralist view may provide the option for balancing individual, community, and systemic interests. Accepting this option would mean proposing ways for drawing boundaries between the community, market, and state economies; defining the rules for each of these economies; re-appropriating the Commons; reducing arms production; re-establishing our organic relationship with nature; revisiting the concept of human development, personal profit, and consumption motives as the guiding principles for decision making; shorter workweek, shared work, universal basic income and healthcare; and decolonizing and demilitarizing the world economy. We need to move away from the unilateral legacies of capitalist and Marxist discourses to a pluralist social discourse, but that would imply taking the position away from political correctness in hindsight to sharing the dreams of the unknown. We might need to move away from the lexicon of freedom, development, and profit to personal fulfilment, nurturing the commons and pluralist forms of democracy. It means taking the risk, becoming controversial, and defining the boundaries of the discourse of consensus.

Organization of the book

Keeping this context in view, contributors to the book have tried to answer the following: (i) Why has the end of Cold War and the collapse of the Soviet Union not led to a peace dividend? (ii) Why is the developed world taking such strong measures to pull out of the open global market that it created after a long and drawn out effort through the WTO regime in 1992? (iii) Why is it profitable and in the interest of the democratic world to produce weapons of mass destruction on a large scale and unprofitable to invest in peace, justice, and social equality? (iv) Why do we witness poverty in the midst of unprecedented abundance? (v) Why did the policy measures taken to correct the market failure, known as a New Deal after the Great Depression, and the creation of a Global Development Assistance Architecture after the end of World War II failed as well, and brought in vogue a neoliberal policy paradigm which in turn failed to address the issues of climate change, social inequality, feminization of poverty and peace and justice, and resulted in fragmentation of global discourse and loss of thought leadership? (vi) Why the global financial architecture that emerged after the end of gold standard in 1974, i.e. acceptance of US dollar as a global reserve currency and creation of money supply through corporate monetary instruments that resulted in the 2007 recession, failed to ensure order, growth, and fairness in the global economy? (vii) Is pursuit of personal gains and profits under the neoliberal economic order in harmony with social interest or does it need to be reimagined to align it with the mitigation of climate change and equitable distribution of gains across social, geographic, and political divides? (viii) What impact has the absence of a meaningful alternative narrative had on the use of media and social media, the creation of alt-facts and the ascendance of emotional truths over the scientific truth? (ix) Can the dreams weaved in the form of the sharing economy, solidarity economy, universal basic income, and other such initiatives find traction in the public square as alternatives to Market and State orthodoxies? Most of the chapters in this volume will look at conventional thinking, norms, and economic, social, and political practices with reference to two key contradictions: social inequality and the planet's survival.

There is no doubt free market only works for a few, and it is imperative that our societies reassert themselves in terms of redistribution and regulation. The issue now is whether we should try and exploit the benefits of markets whilst ensuring they work for everyone (progressive capitalism) or radically transform the structures of our economies through greater economic and social democracy (democratic socialism). In **Chapter 2**, **Susan Spronk** reviews the historical debates about capitalism and development, focusing particularly on the meaning of "the market" in our daily lives and in the shaping of these alternatives. Through case studies, she discusses the fall and rise of socialist experiments in the twentieth century.

The rise of protest movement against growing inequality and precarity over the past decade represents one of the most significant illustrations of the unjust economic order we are living in. This fundamental incompatibility between the current economic system and people's conceptions of "real" democracy is discussed in **Chapter 3** where **Adrian Murray** brings into conversation geographically and temporally distinct episodes of protest and revolution all over the world, drawing together the often-disparate worlds of political economy and everyday struggles for survival in the context of what has long been for many a "diminishing subsistence base."

Income inequality clearly has been on the rise across the globe both within and between countries for several decades. In **Chapter 4**, **Mansoob Murshed** provides a critical evaluation of both the developed and developing world, examining oppressive politics and inequalities in wealth and personal income and employment, emphasizing that while today's rapidly globalized world has far-reaching integrated and interdependent economies, widening income inequality pervasively dampens intergenerational mobility and equity, and heightens global imbalances.

Women are at the intersection of climate change and conflict which are two phenomena with complex relationships and mutually reinforcing adverse impacts. Drawing on examples of civil war, the "war on terror," food security, floods and fires in several parts of the world, **Tricia Glazebrook** in **Chapter 5** analyses the impact of conflict and climate change on gender imbalances which tends to be exacerbated by poverty, while aiming at garnering actions to dissipate such gender biases and enforcing peace, gender equity and a sustainable environment to strengthen women's resilience.

The free market, with its tenets of decentralization and deregulation, as an attack on governments' abilities has contributed to inactions and limited progress in the pursuit for global decarbonization. In **Chapter 6**, **Furqan Asif** asserts that neoliberalism and corporate capitalism stemming from rapid global industrialization have to be transcended to address climate change. Through examples and case studies, he contributes to the argument that the decarbonization of the global economy is possible but will require an optimal mix of government policies, and new practices by corporations, civil society groups, and individuals.

The US dollar has gained much traction over the years and has become the leading global currency relied on by central banks and governments, giving it added legitimacy for domestic and international transactions. In **Chapter 7**, **Abdullah Mamun and Sanni Yaya** present the case of the demise of the dollar. They explore the gradual decline of the dominance of the US dollar in international trade resulting from the rise of competition from the Chinese Yuan, crypto currencies, and the upward spiral in the national debts and protectionist trade policies of the United States.

Globalization has influenced the free flow of capital and labour across national borders and as such migration has been considered an investment which involves both costs and mutual benefits for both labour-sending and labour-receiving countries. **Kazue Takamura** in **Chapter 8** focuses particularly on the case of Asia which remains the world's largest source of international migrants. In examining the unique risks and opportunities embedded in Asia's labour migration regimes, the author throws more light on the concealed intensive migrant surveillance and policing in labour-receiving countries to regulate migrants.

The world's population is projected to stop growing by the end of this century, but Africa remains the only region to grow steadily for the rest of the century. The rapid growth of Africa's population has been associated with its low economic growth, poor population health outcomes, and soaring poverty levels in the region. In **Chapter 9**, **Sanni Yaya**, **Ogochukwu Udenigwe,** and **Helena Yeboah** use the cases of Ethiopia and Nigeria to look at the other side of the debate pointing to the dividend Africa is set to gain at the end of the century due to its population age structure and thus showing that population growth is not always bad for the economic development of countries.

Rasul B. Rais focuses on the fourth industrial revolution and the platform economy. The emergence and proliferation of Information Technology in all aspects of life has brought radical changes in the way business is conducted and how individuals interact. These changes bring the world to the onset of the fourth industrial revolution. While technology increases productivity, efficiency, and cost-effectiveness without compromising quality, its influence on human jobs, consumption and even human dignity in the future may be frightening. In **Chapter 10**, the author scrutinizes the fourth industrial revolution and analyses the concept of distributive justice concerns and their legal challenges.

International development aimed at promoting prosperity and economic growth in the "Third World" and transition nations often arise out of neoliberal and Western practices and ideologies. Their impacts range from the negative outcomes of neoliberal tenets that plagued Africa and Latin America, genocide of Indigenous people to racism and oppression of the "other." **Josh Hawley** and **Nima Hussein** in **Chapter 11** review the discourse of international development as a popular terminology geared towards improving the lives of poor people while arguing that this term which exist to further Western hegemony and advance the interests of colonial and capital states, has undeniably failed to improve the conditions of the poor living in underdeveloped global South.

The Caribbean Sea served as the carrier of fleets that carried wealthy trade that expanded the European empire. In recent years, the control and dominance the United States gained over the Caribbean and the Americas have been threatened by new trade blocks by Venezuela and Cuba, resulting in financial and commercial sanctions by US governments. **Juan Velásquez Atehortúa** in **Chapter 12** analyses the trade blocks, trade wars, and decolonization on the Caribbean Sea.

The Bretton Woods institutions were understood to be an integral part of global stability thereby garnering dominance which was enforced by the financial and military power of the North. **Syed Sajjadur Rahman** in **Chapter 13** examines the radical changes that have been occurring with the twist that the countries and mechanisms that spearheaded the creation and strengthening of these institutions are the same ones posing threats to their impact, legitimacy, and relevance. While a new world order is likely to be created, the author shifts attention to the shape such an order might take and its influence on development financing.

Economic globalization and neoliberal policies have heavily impacted Indigenous peoples as market pressures have resulted in their lands being plundered for capitalist accumulation. For many Indigenous communities, land goes beyond being an economic asset as it provides current and future sustenance and is connected to traditional knowledge, spiritual beliefs, and cultural reproduction which enforces nationhood. In **Chapter 14**, **Carolyn Laude** uses the current treaty negotiations between the First Nation communities and Canadian state to explore contemporary colonial-capitalism in Canada, throwing more light on capitalist expansion at the expense of Indigenous peoples' land. She depicts how a hollow conception of reconciliation wherein state sovereignty supersedes "Aboriginal title" and the public interest supplants Indigenous jurisdiction and rights.

Fayyaz Baqir in **Chapter 15** argues that the grand capitalist order is crumbling, with globalization also failing to promote public participation in political decision making resulting in inequalities of power, resources, and opportunities while increasing intolerance. The emergence of global public opinion has also been met with aggressive counterstrategies on the part of corporations, governments, and inter-governmental organizations leading to alienation and marginalization. The author analyses the significance of the manipulation of public opinion by mass media and corporate finance and the resistance of civil societies in defining new forms of democratic discourse.

References

Ali, T. (2008). Where has all the rage gone? *The Guardian*, 2.

Balch, O. (2013). The relevance of Gandhi in the capitalism debate. The Guardian, 28 Jan. Available at: https://www.theguardian.com/sustainable-business/blog/relevance-gandhi-capitalism-debate-rajni-bakshi.

Bauzon, K. (2016). Themes from the history of capitalism to the rise of US Empire in the Pacific, with Annotations from selected works of E. San Juan, Jr. *Kritika Kultura*, *0*(26), 408–443. doi: http://dx.doi.org/10.13185/KK2016.02623.

Bockman, Johanna. (2013). Neoliberalism– Contexts, August 11.

Bentham, Jeremy. (1823). *An Introduction to the Principles of Morals and Legislation*. London: Clarendon Press, 1907. Available at: http://www.econlib.org/library/Bentham/bnthPML.html.

Brosio, R. A. (2008). Marxist thought: Still primus inter pares for understanding and opposing the capitalist system. *Journal for Critical Education Policy Studies*, v6 (n1). May 2008.

Chang, H. J. (2012). *23 Things They Don't Tell You About Capitalism*. New York: Bloomsbury Press.

Chavance, Bernard. (2003). The historical conflict of socialism and capitalism, and the post-socialist transformation. In: John Toye (ed.), *Trade and Development*. Edward Elgar Publishing.

Christianto, Haryadi. (2018). Neoliberalism: Past, present…future? Canadian International Council, Aug 10, 2018 https://thecic.org/neoliberalism-past-present-future/.

Frieden, Jeffry. (2012). The modern capitalist world economy: A historical overview. In: Dennis Mueller (ed.) *Oxford Handbook of Capitalism*. New York: Oxford University Press.

Graeber, David. (2019). *Against economics, The New York Review of Books*, December 5, 2019 issue.

Hegel, G. W. F, Baillie, J. B and Lichtheim, G. (1967). *The Phenomenology of Mind*. New York: Harper & Row.

Hanauer, Nick. (2019). Founder of the public-policy incubator civic ventures. The Atlantic July 2019 issue.

Iber, Patrick. (2018). Worlds apart: How neoliberalism shapes the global economy and limits the power of democracies. The New Republic April 23, 2018.

Rodrigues, João, Kuttner, Robert. (2019). Can democracy survive global capitalism? *Critical Review of Social Sciences*, 119, pp. 201–204.

Kocka, Jürgen. (2010). Writing the History of Capitalism. First Gerald d. Feldman memorial lecture delivered at the GHI Washington, April 29, 2010. Available at: https://docplayer.net/41462644-Writing-the-history-of-capitalism-jurgen-kocka.html.

Kuttner, Robert. (2019). Neoliberalism: Political Success, economic failure, The American Prospect, June 25, 2019.

Kahneman, D. (2011). *Thinking, Fast and Slow.* New York: Farrar, Straus and Giroux.

Linder, Marc and Sensat, Julius. (1977). *The Anti-Samuelson.* Volume 1. Urizen Books New York: London.

Mann, Geoff. (2020). The inequality engine. *London Review of Books,* 42(11).

McKenzie, R. (1981). The necessary normative context of positive economics. *Journal of Economic Issues,* 15(3), pp. 703–719. Retrieved June 22, 2020, from www.jstor.org/stable/4225070.

Moyn, Samuel and Wertheim Stephen. (2019). The infinity war. Washington Post, December 13, 2019.

Plato, Jowett, B. and Greene, W. C. (1927). *The Dialogues of Plato.* New York: Boni & Liveright.

Sachs, Jeffrey D. (2011). The Global Economy's Corporate Crime Wave Project Syndicate. Apr 30, 2011. Available at: https://www.project-syndicate.org/commentary/the-global-economy-s-corporate-crime-wave?barrier=accesspaylog.

Scott, Bruce R. (2006). *"The Political Economy of Capitalism"* Harvard Business School Working Paper, No. 07-037, December 2006.

Schumpeter, J. A. (2008). *Capitalism, Socialism, and Democracy.* New York: Harper Perennial Modern Thought.

Sen, Amartya. (2000). *Development as Freedom.* New York: Anchor Books

Skidelsky, Robert. (2019). *Money and Government: The Past and Future of Economics.*

Smith, Adams (2018) *An Inquiry into Nature and Causes of Wealth of Nations,* Volume 1 Google Books 18 October 2018

Stiglitz, Joseph E. (2019). Three decades of neoliberal policies have decimated the middle class, our economy, and our democracy. *Market Watch* May 13, 2019. Available at: https://www.marketwatch.com/story/three-decades-of-neoliberal-policies-have-decimated-the-middle-class-our-economy-and-our-democracy-2019-05-13.

Taleb, N. N. (2010). *The Black Swan: The Impact of the Highly Improbable.* 2nd ed., New York: Random House.

Vidal, Gore. (2005). *Imperial America: Reflections on the United States of Amnesia.* New York: Nation Book.

2

DEVELOPMENT ALTERNATIVES BEYOND THE "FREE MARKET"

Progressive capitalism and democratic socialism

Susan Spronk

> *"The economic anarchy of capitalist society as it exists today is, in my opinion, the real source of the evil. ... I am convinced there is only one way to eliminate these grave evils, namely through the establishment of a socialist economy"*
>
> Albert Einstein (1949/2009)

Introduction

Two decades into the twentieth century, it is becoming increasingly clear to more people that capitalism—at least in its neoliberal form—has failed. Growing uncertainty, privatization and deepening financialization have widened inequality, increased precarity, and intensified poverty. These challenges are magnified by a climate emergency which has made procuring what we need for daily life on earth more challenging for the majority of humanity, and displaced millions of people threatened by extreme weather events such as hurricanes, droughts, and floods. Those least at fault for this situation are the ones who pay the highest price.

The pandemic of COVID-19 has exposed and exacerbated these inequalities, shining a light into the cracks in the foundations of the racist, patriarchal society that colonialism and capitalism have built. Yet it has also helped to create the conditions where the impossible suddenly seems possible. Governments the world over suddenly found room in their budgets to provide basic income supports and housing. Workers who were once considered to be low-skilled are now considered "essential."

More and more people are asking the questions; why do poor have no food, decent work, housing, or access to basic services such as health, education, and recreational opportunities. This era of protest and revolt provides an opportunity to think about how to rebuild. As Roy (2020 recently commented, the pandemic is a "portal." But what is it that we want to rebuild exactly? Is it capitalism or neoliberal globalization that is the problem, and what are the alternatives? This chapter helps to answer these questions by outlining the basic flaws of capitalism, and comparing democratic socialism and progressive capitalism, two concepts that have dominated the debate on alternative economic systems in North America, especially the United States (US).

The chapter is organized as follows. In the first section, we discuss the limits of capitalism, particularly in its neoliberal phase, and explain why more and more people have come to seek alternatives after the financial crisis of 2008. In the second section, we define the two concepts democratic socialism and progressive capitalism, arguing that these are closely related and ought to be understood as concepts along a continuum of social democracy.

Limits of capitalism

Capitalism is a social order in which productive wealth is privately owned and is deployed solely to make profit. Capitalists (the owners of capital) do not invest their money to meet human need. If they meet a human need (whether that need is "real" or "manufactured"), it is always second to their principal goal of making a profit. As Marx (1993, p. 352) writes, capitalist production, therefore, "should never be depicted as something that it is not, i.e. as production whose immediate purpose is consumption. [...] This would be to ignore completely its specific character." This tendency to focus on producing for profit rather than human need gives rise to one of its inherent contradictions: overproduction.

Overproduction arises under capitalism because the unlimited drive to expand production periodically comes into collision with the limited confines of the market economy. Plenty of people want and need things, but do not have the money to buy them. According to economists, they lack "effective demand." This strange phenomenon of overproduction arises when excess commodities, goods produced for sale, cannot be sold. Of course, this situation is related to the fact that the working class cannot buy back the full value of what it produces (sometimes identified as "under consumption"), but the lack of purchasing power of the working class is only one side of the equation. After all, there is nothing new about the restriction of the consumption of the labouring masses. Overproduction, on the other hand, is a phenomenon that is unique to capitalism, and it is linked to its relentless drive for profit. This dialectical contradiction lies at the heart of the capitalist system. It is one of the reasons that capitalism has a tendency towards crisis.

The capitalist system tries to overcome the problem of overproduction in two ways. First, as David Harvey (1982) has observed, capitalism perpetually seeks out "spatial fixes" for its problems. As we have seen since the birth of capitalism, capitalists constantly seek to move production to new zones in the world economy to benefit from low-wage zones where workers are unorganized and willing to work for cheap since they have few choices. Second, capitalists are always seeking new markets, by absorbing non-capitalist markets into its orbit. These dynamics are fundamental to understanding the concepts of "development" and "globalization" that define the two major eras of capitalist expansion since the Second World War.

The development era, aka the "Golden Age" of capitalism, 1945–1970s

There is a tendency amongst intellectuals born before the 1980s (that is, most of your professors) to sentimentalize the 1950s and 1970s as the "Golden Age" of American and Western capitalism. During a period of about two decades, there was an unprecedented cycle of economic growth related to the expansion of the world market. United States, Western European and East Asian countries in particular experienced unusually high and sustained growth, together with very low rates of unemployment. As Paul Street (2015) argues, it was a time when economic inequality declined, significant Civil Rights victories were achieved, unemployment was low, the middle class expanded, wages and consumption rose, the welfare state grew, college was more affordable, and there was a flourishing of counter-culture and popular music, and great achievements in science and technology. Political leaders felt optimistic after the triumph of the Allied Powers in the Second World War, hoping that the joys of liberal democracy and capitalism would be spread across the globe.

The economic model that emerged at this time is known as "embedded liberalism"—a form of capitalism that was embedded in society, constrained by political concerns, and devoted to social welfare. Embedded liberalism sought to exchange a decent family wage for a docile, productive, middle-class workforce that would have the means to consume a mass-produced set of basic commodities. These principles were widely applied after World War II in the North America and Europe. Policymakers believed that they could use Keynesian principles to ensure economic stability and social welfare around the world, and thus prevent another world war. They developed the Bretton Woods Institutions (which would later become the World Bank, the IMF, and the WTO) toward this end, in order to smooth out balance of payment problems and to foster reconstruction and development in war-torn Europe.

Trade unions were also a key social actor during this period. In the aftermath of major strike waves in North America at the end of WWII, governments recognized trade unions and guaranteed that any increases to labour productivity would be shared with workers through increases to wages and benefits in collective bargaining in what is known as the "post-war compromise." In return, trade union leaders played a key role in disciplining the workforce, forbidding strikes unless legally mandated, that is, only if the collective agreement had expired. In addition to favourable labour laws that made it easier to organize trade unions, union density grew across North America and Europe. As key social actors, unions exerted pressure on political leaders to expand the welfare state. These efforts resulted in the mixed economy in the United States and the United Kingdom, and the more social-democratic welfare states in Europe and Canada.

At the time, the social democratic compromise appeared to be the best "middle way" between the unbridled capitalism that caused the Great Depression of the 1920s and the increasingly totalitarian Stalinist model of the economy that characterized the Soviet Union and its satellites. In the context of the Cold War "development" was a project that sought to integrate non-aligned states into the orbit of the western capitalist nations to stave off the threat of communism. At this time in world history, communism was capitalism's greatest threat. In the early 1950s in the United States, anti-communism reached a fever pitch when Americans were dragged in front of a committee chaired by Senator McCarthy. Those found to be communists lost their jobs, had their careers destroyed, or were put in jail. That is, while there is little doubt that this "Golden Age" of capitalism created enormous wealth and fostered technological innovation, it also had a dark underside.

The Golden Age of capitalism are also the years of prime hegemony of the United States, which has played the role as the global police for capitalism. While leaders of the "free" world pay lip service about the importance of liberal democracy, time and again, elected leaders who threatened Western business interests with nationalist, redistributive policies—Patrice Lumumba in the Congo, Thomas Sankara in Burkina Faso, Salvador Allende in Chile—were deposed and replaced with kleptocratic regimes more amenable to imperial interests often by violent military invasion. In Asia, those who have dared to advance alternative economic systems in countries such as in Vietnam were crushed by brutal military invasion. By the late 1970s, 17 countries of 20 Latin American countries were controlled by military-authoritarian governments, backed by the US. It is not by accident or the outcome of bad luck of geography that explains why the former colonies remain poor; these processes of dispossession, theft, plunder, and military invasion have *created* the Third World.

It is also important to remember that even in the capitalist core, the benefits of growth have never been distributed evenly, especially along racial lines. To cite one example, urban development policies of the New Deal era were overtly racist. The Federal Housing Administration (FHA), which was established in 1934, refused to insure mortgages in and near African-American neighbourhoods—a policy known as "redlining." At the same time, the FHA was subsidizing builders who were mass-producing entire subdivisions for whites—with the requirement that none of the homes be sold to African-Americans. In Canada, the gains of the "Golden Age" have never been shared with the First Nations, Métis, or Inuit peoples who were dispossessed of their lands and territory. Until 1960, Indigenous people with status were not even allowed to vote.

Finally, the form of industrial development involving the mass production of cheap commodities has pushed production beyond ecological limits. Researchers Rockstrom et al. and Steffen et al., for example, have estimated that five out of seven planetary boundaries have already been exceeded, and the remaining two are two-thirds of the way (cited in Hickel 2019). Research on the ecological footprint has estimated that we would need 5 planet Earths if every person on the planet consumed at the rate of the average US citizen. The relentless drive for economic growth may be the Achilles' heel of the capitalist system, the ecological devastation of which was not predicted by Marx himself.

The globalization era: neoliberalism and the unravelling of the post-war compromise, 1970s-

The post-war compromise that underpinned the "Golden Age" began to erode in the 1970s, when capitalists in the global North faced a crisis of capital accumulation marked by a combination of rising unemployment and accelerating inflation. The patterns of counter-cyclical spending that were inspired by Keynes no longer appeared to work in the turbulent years of the 1970s, due to an unprecedented situation of high unemployment and inflation known as "stagflation."

During the heady days of the expansion of the world economy, the economic power of the upper classes could be restrained and labour could be accorded a much larger share of the economic pie. Elites considered this deal acceptable when the pie was growing but when the pie began to shrink, they felt threatened and went on the offensive. For this reason, as French economists Dumenil and Levy have stressed, neoliberalism was "from the very beginning a project to achieve the restoration of class power" (cited by Harvey 2005, p.16).

In the Oxford English Dictionary, neoliberalism is defined as an economic ideology, a "modified or revived form of traditional liberalism, especially one based on belief in free market capitalism and the rights of the individual." Neoliberalism is usually associated with a set of policies that were introduced under the "Washington Consensus" of the 1980s, including the deregulation of financial markets (particularly removing controls over capital flows), currency devaluation, the dismantling of tariff barriers, the implementation of social austerity such as cutting subsidies on basic foodstuffs, and privatization. However, as Susanne Soederberg has warned, the definition of neoliberalism should not be limited to a policy framework alone: "neoliberalism is not about less state intervention engendered by the faceless forces of globalisation, but, instead, a new form of capitalist political domination propelled by class struggle" (Soederberg 2001, p. 63).

The first experiment in neoliberalism was in Chile after the Pinochet coup of September 11, 1973, which was backed by US corporations, the CIA, and US Secretary of State Henry Kissinger. Pinochet violently suppressed the left and popular organizations, especially trade unions. The CIA also funded the installation of conservative US-trained economists known as "the Chicago boys" eager to implement free-market doctrine. The outcome was very successful for the corporate rich and their military friends, but it was a predictable disaster for most Chileans. Chile became one of the world's most unequal societies and entered a long period of brutal dictatorship.

The next milestone in the onset of the neoliberal era was the election of Thatcher in UK in 1979 and Reagan in the US in 1980. Under Thatcher's rule, the government dismantled state-owned enterprises and broke the miner's strike. Similar to Thatcher, one of Reagan's first moves in office was to break the PATCO strike of air traffic controllers. This strike was particularly significant since it was a union of white-collar professionals. It sent shock waves throughout the union movement. Wage levels have since dropped. At the same time, corporate taxes were reduced dramatically, and the top personal tax rate was reduced from 70 to 28% in what was billed as the "largest tax cut in history" (Harvey, 2005, p. 26).

A third milestone in the neoliberal era is the Third World Debt crisis. After 1973, corporations, particularly private banks, were looking for a place to invest the "petrodollars" floating around in the world economy. They became more focused on lending to foreign governments, many of them authoritarian. Eventually, these private lenders become over-exposed and afraid of default, especially in the context of the contraction of the world economy. Just before the election of Reagan, the US hiked interest rates in an attempt to control the spiralling inflation. In a very short period from 1979 to 1981, interest rates skyrocketed, going from negative (due to inflation) up to 20%. The interest rate hike created balance of payments crises in the countries that had borrowed massive amounts of money from both multilateral lenders and private banks.

The solution to this crisis was neoliberal Structural Adjustment Programs, orchestrated by the International Monetary Fund and the World Bank. While these programs ostensibly aimed to stabilize local economies to get the economic fundamentals right, the real intent was to get the lenders their money back by prioritizing export. Under Structural Adjustment Packages, new loans would be granted to solve debt repayment problems, but on the condition that recipients geared their economies to neoliberal policies intended to remove impediments to business and foreign investment and trade: eliminate tariffs and other protections for local industries, sell national assets (to rich multinational corporations, at fire-sale prices), eliminate subsidies of basic foodstuffs, devalue the currency (making your exports cheaper for the rich world to buy, and making you pay more for your imports from it), free up access to resources for foreign corporations, cut wages and welfare, and allocate national income to paying back debt. In other words, the aim of neoliberal SAPs was to totally reverse the previous national "developmentalist" policies and redirect national income and assets to debt repayment. As David Harvey (2005) writes, debt has become the foremost weapon of this "new" imperialism:

> From the 1950s through the 1970s, Western powers had struggled to prevent the rise of developmentalism in the South. What they failed to accomplish through piecemeal coups and covert intervention, the debt crisis did for them in one fell swoop.... Debt became a powerful mechanism for pushing neoliberalism around the world (p. 155–156.)

As Philip McMichael (2016) describes, the IMF and the World Bank played a key role in the crisis as the "managers of global capitalism." The IMF forced through privatization of more than \$2 trillion of assets in developing countries between 1984 and 2012 (Hickel 2017, p. 171.)

The collapse of the Berlin Wall in 1989 and the dissolution of the Soviet Union in 1991 represent another milestone in the history of neoliberal globalization. The defeat of communism and the end of the Cold War was heralded as the "end of history" (Fukuyama 1989) and was supposed to bring another golden age for Western, capitalist-driven liberalism. As more and more countries democratized optimists hoped that humanity would turn to post-ideological goals, like ending hunger. Alas, that is not what happened.

Financialization and the Great Recession of 2007–9

While neoliberalism may be about the restoration of class power, it has not necessarily meant the restoration of economic power to the same people. Financialization refers to the way that the financial parts of the economy (e.g. finance, insurance, and real estate), have grown far faster than the other of the parts of the "real" economy, especially in the UK and the US. As Harvey quips, "Neoliberalism, in short, has meant the financialization of everything" (2005, p. 33).

The financial sector received a boost in 1971 when Nixon unpegged the US dollar from the gold standard, thus terminating the Bretton Woods System, and opening up new market: a "global casino" in which investors speculate on currency. Financialization was further accelerated by financial deregulations during Reagan's presidency and the accompanying tax cuts, which favoured the wealthy. The collapse of the Bretton Woods system and the deregulation of finance help to explain why the system has become more unequal, but also more unstable. In this new financialized economy, the CEOs of banks and financial corporations have done particularly well, but so have corporations taking advantage of new markets overseas.

Financialization has made global capitalism more prone to crisis. Remarkably, as Wolf (2008, p. 31) observes, between 1945 and 1971 there were only 38 financial crises, compared to the early neoliberal period (1973–1997) when there were 139. While there have been many financial crises that have slowly eroded the hegemony of neoliberalism since 1997 such as the Asian Flu, the Great Recession of 2007–9 is arguably the most important crisis that rocked the core of the world economy with truly global effects.

In his modern classic, *Capital in the Twenty-First Century*, the French economist Piketty (2014) suggested that growing inequality in America contributed directly to the country's financial instability. One consequence of increasing inequality, he wrote, "was virtual stagnation of the purchasing power of the lower and middle classes in the United States, which inevitably made it more likely that modest households would take on debt, especially since unscrupulous banks and financial institutions, freed from regulation and eager to earn good yields on the enormous savings injected into the system by the well-to-do, offered credit on increasingly generous terms" (2014, p. 279). A decade after the crisis, income inequality is the highest it has been in the United States since it started being tracked over five decades ago. Disparities in wealth are even more extreme. Meanwhile, household debt has exploded, driven largely by student loans and credit card debt, which steadily grow as wages stagnate and jobs have become more precarious. These trends disproportionately affect young people, although that has not stopped the financial class from blaming them for the "sluggish economy."

The effects of the Great Recession were not just economic, but also social and political. The crisis was arguably the most significant event of the twenty-first century (before the COVID-19 crisis), and the largest single economic downturn since the Great Depression. The years after the crisis saw sharp increases in polit-ical polarization and the rise of populist movements on both the left and right in Europe and North America, culminating in Brexit in the UK and the election of Donald Trump in the US. The rise of right-wing populism was not just a response to the financial crisis and its painful consequences; it was a response to the fact that nothing fundamentally changed in its aftermath. The big banks remained too big to fail, executives who had overseen rampant fraud remained free (with their generous bonuses intact), income and wealth inequality contin-ued to grow out of control, and wages continued to stagnate as billionaires saw their wealth multiply. In other words, the economy "recovered" for those on top, while the recession lingered for everyone else.

It is no small wonder that within the contemporary context of compounding crises, more and more young people are seeking alternatives. The American left currently finds itself on unfamiliar political terrain. Interest in socialism is grow-ing, especially among a younger generation. Outrage toward Trump's racism and xenophobia, millennials' anxieties about their economic prospects, and a deep-ening scepticism about the ability of establishment to address these problems has caused many to seek answers. The American left hasn't experienced such a rapid influx of activists and adherents since the 1960s, with organized movements such as Occupy Wall Street and Black Lives Matter capturing the public imagination. In the US, seventy years after being actively repressed by McCarthy, the term socialism is back on the political agenda. Young people no longer associate the term "socialism" with the authoritarian and defunct states of the former Soviet Union, but with Bernie Sanders.

Varieties of social democratic capitalism: Progressive capitalism and democratic socialism

The debate between progressive capitalism and democratic socialism unfolded during the electoral contest to decide who would be the candidate to run against President Donald Trump in the 2020 election for the Democratic Party. While presented as two different camps, democratic socialism in its contemporary usage in the US has a lot in common with progressive capitalism. These two con-cepts are best thought of as lying along a continuum rather than two opposite poles. Their proponents favour slightly different policy instruments that aim to pursue a program of distributive justice, using resources generated by a capitalist economy founded on private property relations. Neither project proposes true socialism (communism) but rather different forms of social democracy that differ in terms of the degree of state involvement in the market with respect to the amount of spending on infrastructure and service delivery, and the ambitions of their tax reform projects.

Progressive capitalism is a relatively new term embraced by left-of-centre economists—most notably Joseph Stiglitz (2020)—and politicians such as Senator Elizabeth Warren, who was a front-runner in the Democratic primary in 2019. Democratic socialism, on the other hand, is a term that is most closely associated with Senator Bernie Sanders, who faced off in the Democratic primary race against Hillary Clinton in 2016 and Elizabeth Warren in 2019. Since democratic socialism has been the topic of intense debate for over a century, it is impossible to name just one author or even one group of authors associated with the term.

Winner of the Nobel-Memorial Prize for Economics in 2001 and the former senior vice president and chief economist of the World Bank, Stiglitz has long argued that neoliberal globalization has worked for a select few rather than the masses (see his 2002 book *Globalization and its Discontents*, Stiglitz 2002). In his most recent book, *People, Power and Profit* (Stiglitz 2020), he argues that the only way to save capitalism from itself is for governments to recognize the central role that the state plays in setting the rules of market. Public policy must make markets serve society rather than the other way around. The first part of the book examines four trends that Stiglitz believes to have set the US on a path to a poorly performing, highly unequal economy: monopoly power, mismanaged globalization, weak to non-existent financial regulation, and the heavy reliance on new technologies that enable further exploitation and psychological manipulation. The second part moves from diagnosis to cure. In terms of economic policy, Stiglitz supports a carbon tax, more infrastructure spending, government job guarantees, a financial transactions tax, expanding Social Security benefits, family leave policies, a higher minimum wage, free public college education, and government-provided childcare.

Sander's democratic socialism proposes a similar mix of policy responses, aiming to create a form of mixed economy with a slightly heavier role of the state in providing public goods such as health care and education. This form of socialism resembles the stronger social democracies in the global North like the Scandinavian countries (i.e. Sweden, Norway, Denmark, and Finland). As Sanders put it, the ideal is "an economy in which you have wealth being created by the private sector, but you have a fair distribution of that wealth, and you make sure the most vulnerable people in this country are doing well" (cited in Bolton 2020, p. 335).

In terms of their 2020 policy platforms, Sanders and Warren had a lot in common. Both candidates aimed to ban private schools, forgive student debt, and make university and college free, introduce legislation to break up the big banks and agribusiness, impose wealth taxes on the richest Americans, impose a carbon tax, and reform the criminal justice system including the end of private prisons, minimum sentencing, and bail.

There were a few differences in which Sanders stood out as the more radical candidate when compared to Warren. These included his plans to eliminate tax breaks for "offshoring" (Warren's plan promises to track illicit flows but does not say what her government would do about them) and his massive national housing plan (Sanders' plan was estimated to cost $2.5 trillion; Warren's $470 billion). And, crucially, while both Warren and Sanders endorsed plans to create a universal, single-user pay healthcare system known as "Medicare for All," during the campaign Warren backed off initial promises to implement the plan immediately, promising to transition to the system within three years of her election.

The key difference between Warren's democratic capitalism and Sanders's democratic socialism lies not so much in how these academics and politicians define them, but in what they mean to their supporters. There is no social movement organization that articulates around progressive capitalism *per se*. Democratic socialism, on the other hand, is a concept endorsed by social movements. The membership of the Democratic Socialists of America, which threw its weight behind Bernie Sanders, has skyrocketed since 2015, growing from 6,000 to 70,000 members (as of June 2020). Democratic socialism is inspiring excitement, particularly amongst younger voters. Many supporters of Sanders believe that democratic socialism is something that could potentially break with capitalism, while supporters of Warren seek to reform capitalism. Warren emphasizes that she is "capitalist to [her] bones" (cited by Farrell 2019).

Although Sanders ultimately failed to win the nomination, many consider his campaign a success. It helped to build a social movement that is still growing, pushing the policy debate forward to expand the political imagination of many Americans, who support key policies that would reduce poverty and improve the living standards of the majority rather than the few, such as Medicare for All, a Green New Deal, and free college tuition. Largely thanks to Bernie, the term "socialism" is again part of the political vocabulary in the US.

Democratic socialism and the socialist horizon

Over time, the meaning of social democracy and socialism have shifted. Sander's plan may look radical now, but in terms of spending and state intervention in the economy, it is not as ambitious as Roosevelt's New Deal. While democratic socialists in the US aim to humanize capitalism by creating a welfare state to try to build capacity for collective struggle, they also aim to go beyond social democracy by building socialism.

Democratic socialists seek to break with capitalism to create a more participatory and egalitarian political *and* economic system (Bolton 2020). They argue that true democracy is not possible within capitalism. On this view, social democracy is not the goal in and of itself, but a *transitional stage* on the way to something else. The longer-term goal is to build a new form of society based upon an economic system that collectivizes ownership. It is not enough to try to democratize capitalism by sharing the wealth more equitably. The production system requires a radical overhaul.

The Meidner Plan in Sweden is one often-cited example of a project within social democracy that hoped to create socialism by buying out private capital and replacing it with a form of collectivized ownership. In 1976, the Swedish Trade Union Confederation proposed the Meidner Plan, which aimed at the gradual socialization of Swedish companies through a wage earner fund. The idea was that private producers would slowly be replaced within two generations by transferring company shares to the people who worked there through "wage earner funds." The Social Democratic Party lost the elections in 1976, returning to government only in 1983, and the Prime Minister who most supported the plan, Olof Palme, was tragically assassinated in 1986. Similar to other social democratic parties of the "Third Way," the Swedish Social Democratic Party drifted towards neoliberalism in the context of the globalization of production, trade and finance. Activists and researchers in different parts of the world still take inspiration from the Swedish Meidner Plan. Left-wing think tanks in the US and UK, such as the Next System Project, the Democracy Collaborative and the Commonwealth, have put forward policy ideas that helped to inform Sander's campaign promise to create an employee ownership plan. In the 2019 campaign, UK Labour leader Jeremy Corbyn also proposed a plan to force companies with more than 250 employees to create Inclusive Ownership Funds, along the lines of the Meidner Plan.

While democratic socialists argue that social democracy might be better than a neoliberal form of capitalism since it raises the living standards for a larger number of people, as a system of production and consumption, the private property at the base of the production system confronts immediate limits. Furthermore, key social democratic reforms, such as raising the level of taxes on the wealthiest individuals and corporations may expand the offer of public goods and essential services, but they leave the capitalist system of production untouched, doing nothing to change economic decision-making. Private individuals and firms still decide what is produced, how it is produced, how much it is sold for, and how much workers are paid. Social democracy is therefore limited because it does not question the way that wealth is created in capitalist society, in which workers are only paid a fraction of the value they produce. It remains a system that is organized around producing for profit rather than human need. Mike McCarthy (2018) clarifies:

> Democratic socialism… should involve public ownership over the *vast majority* of the productive assets of society, the elimination of the fact that workers are forced into the labour market to work for those who privately own those productive assets, and stronger democratic institutions not just within the state, but within workplaces and communities as well. Our characterization of democratic socialism represents a profound deepening of democracy in the economy.

According to the democratic socialist critique, social democracy is limited because the production system remains unchanged. Governments and states remain vulnerable to the pressures and limitations of competition in the world market. Indeed, social democracies have slowly shifted towards more neoliberal forms of capitalism due to the pressure exerted on producers within the world market. In one of the more masterful surveys that compares "models of capitalism" in the advanced industrialized countries of Europe, North America and Asia, David Coates (1999) argues that most writers tend to place too much emphasis on variety, and too little on the homogenizing effects of capital. He observes that national capitalisms are not self-contained and self-sufficient; they are interconnected cross-nationally. He argues that we therefore need to understand the way that neoliberal globalization has altered the dynamics of international finance, trade and production. He observes that "state-led" and "negotiated" or "consensual" capitalisms, such as those in social democratic countries in which the state played a central role in creating a welfare state maintained through a high level of taxation, were once more successful than those he terms "market-led."

Conclusion

At base, many debates in international development are about how to address poverty, promote human rights, and create economic and social alternatives in which all the world's people not only meet their basic needs but flourish. This chapter argues that while many scholars of an older generation remain nostalgic about the era of "embedded liberalism" of a by-gone era, capitalism has inherent contradictions that cannot be solved through regulation alone. Rather, socialist models of development offer a potential solution to the problems of development resolving the conflict between the owners of capital, workers, and consumers by transforming the capitalist-owned economy into a socially-owned economy. Rather than profits for a few, the goal is to achieve social well-being of all. Even Francis Fukuyama, who once heralded the triumph of liberal capitalism as the "end of history," (1989) is calling for the kinds of social protections that were once associated with socialism (cited by Eaton 2018).

This chapter also reviewed the recent debate on the US Left on progressive capitalism and democratic socialism, arguing that these concepts are best thought of different points along a continuum, that is, varieties of social democracy. We have also argued that Bernie Sander's campaign has inspired a nascent socialist movement in the US, which takes inspiration from socialist experiments of the past such as the Meidner Plan in Sweden, which was a peaceful way to transition from a social democratic to a socialist economy.

Lastly, while the failure of various forms of capitalism has not been equated with the failure of capitalism in the public mind, the failure of one form of socialism (Stalinism) has been presented as the failure of socialism. This chapter argues, by contrast, that the underlying failure of various forms of capitalism is the inherent conflict between equity, justice, and survival of the planet and profit-driven growth.

References

Bolton, M. (2020). Democratic Socialism' and the concept of (Post) capitalism. *The Political Quarterly*, 91(2), pp. 334–342.

Busky, D. F. (2000). *Democratic Socialism: A Global Survey*. Westport, CT: Praeger Publishers.

Clement, W. (1994). Exploring the limits of social democracy: Regime change in Sweden. *Studies in Political Economy*, 44, pp. 95–123.

Coates, D. (1999). *Models of Capitalism: Growth and Stagnation in the Modern Era*. London: Polity Press.

Eaton, G. (2018). *Francis Fukuyama interview: "Socialism ought to come back." New Statesman.* October 17. Available at: https://www.newstatesman.com/culture/observations/2018/10/francis-fukuyama-interview-socialism-ought-come-back?fbclid=IwAR1-4Z3Xc GqIt6KJHAuC_s7_65EOHqwczjjye22AfUhBiIXk6_GC1IneZUc

Einstein, A. (1949/2009). "Why socialism?" *Monthly Review*. Vol. 61, Issue 1 (electronic version). Available at: https://monthlyreview.org/2009/05/01/why-socialism

Farrell, H. (2019). Socialists will never understand Elizabeth Warren. *Foreign Policy*. December 12. Available at: https://foreignpolicy.com/2019/12/12/elizabeth-socialist-understand-capitalism-pro-market-leftist/.

Fukuyama, F. (1989). The End of History? *The National Interest*. Summer.

Harvey, D. (1982). *Limits to Capital*. Oxford, UK: Basil Blackwell Publisher.

Harvey, D. (1992). *Condition of Post-Modernity: An Enquiry into the Origins of Cultural Change*. Cambridge, MA: Harvard University Press.

Harvey, D. (2005). *A Brief History of Neoliberalism*. London: Oxford University Press.

Hickel, J. (2017) *The Divide*. London: Penguin Random House.

Hickel, J. (2019). The contradiction of the sustainable development goals: Growth versus ecology on a finite planet. *Sustainable Development*. 27, pp. 873– 884. https://doi.org/10.1002/sd.1947

Marx, K. (1993). *Capital: A Critique of Political Economy: Volume III*. Trans. D. Fernbach. New York: Random House Press.

McCarthy, M. (2018). 'Democratic socialism Isn't social Democracy' *Jacobin*. July 8. Available at: https://jacobinmag.com/2018/08/democratic-socialism-social-democracy-nordic-countries

McMichael, P. (2016). *Development and Social Change. Sixth Edition*. Thousand Oaks, CA: Sage Publications.

Piketty, T. (2014). *Capital in the 21st Century*. Trans. A. Goldhammer. Cambridge, MA: Harvard University Press.

Roy, A. (2020). The pandemic is a portal. *Financial Times*. April 3, 2020.

Soederberg (2001). State, Crisis, and Capital Accumulation in Mexico. *Historical Materialism*.9(1):61–84, DOI: 10.1163/156920601760039186

Stiglitz, J. (2002). *Globalization and Its Discontents*. New York: W.W. Norton & Company.

Stiglitz, J. (2020). *People, Power and Profits*. New York: Penguin and Random House.

Street, P. (2015). "The Not-So Golden Age: a Radical and Eco-Socialist Take on Post-WW II America and 'the Anthropocene'" *Counterpunch*. October 16. Available at: https://www.counterpunch.org/2015/10/16/the-not-so-golden-age-a-radical-and-eco-socialist-take-on-post-ww-ii-america-and-the-anthropocene/.

Wolf, M. (2008). *Fixing Global Finance*. Baltimore, MD: Johns Hopkins University Press.

3

SOCIAL REPRODUCTION, SOCIAL MOVEMENTS AND MARKET FAILURE

Adrian Murray

Introduction

In a dramatic shift away from the state-managed capitalism of the post-World War II era, production and consumption have been transformed over the last four decades in the period widely known as neoliberal or financialized capitalism. In the realm of the latter, welfare state entitlements—as thin or non-existent as they were in many countries, especially in the South—have been continuously undermined and replaced with shifting versions of increasingly market mediated access to formerly public services and other aspects of the "social wage." Whether through public-private partnerships, or outright privatization, global development institutions have remained thoroughly committed to the market-based restructuring of these goods and services essential to producing and reproducing human life (World Bank 2003; 2015). This constantly evolving best practice has, however, proved woefully inadequate in meeting the needs of the majority of the global working class evidenced by the series of dramatic market failures suffered by privatized services over the last several decades (Bakker 2010; Kishimoto and Petitjean 2017; McDonald 2015). For example, despite the decades-long commitment on the part of global development institutions to this market-based agenda, the World Health Organization estimates that 2.1 billion people still lack access to improved water and 4 billion to improved sanitation (UNICEF and WHO 2019).

This privatization agenda did not proceed without opposition however, but rather was and continues to be vigorously contested by those bearing its brunt (McDonald 2016). From the IMF riots of the 1970s and 1980s, to the anti-globalization protests in the North and South at the turn of the millennium, to protests against austerity following the Great Recession which have continued for the last decade, popular protest has always been closely linked with the everyday living conditions of working-class people. Movements composed of those predominately outside of the political class and formal worker's organizations have played a key, even leading role in virtually every case of social organization and mobilization around access to the basic necessities of life in this period, much as they did prior.

Informed by Social Reproduction Theory and Analysis (SRT), this chapter brings into conversation these and other geographically and temporally distinct episodes of protest around the labour and access to the means of social reproduction—"all those... processes involved in maintaining and producing people... and their labour power on a daily and generational basis" (Luxton and Bezanson 2006, p. 3). This includes the labour predominately done in the home to feed, clothe, educate, socialize, nurse, etc. children, adults, the elderly, and others in need of care, as well as in the formal and informal neighbourhood, community, and state institutions at a variety of scales involved in these processes of "life-making" (Bhattacharya 2020b). And here we are talking about all manner of public services, housing, transport, parks and recreation facilities, daycares, schools, after-school programs, hospitals, care homes, and so on. Following in this tradition, this chapter draws together the often-disparate worlds of political economy and everyday survival/life in the context of what has long been for many a "diminishing subsistence base" (Nash 1994, p. 10), to reveal the foundations of these political confrontations in the contradictions of the capitalist mode of production, and the class struggle at its centre.

This chapter is organized as follows. The first section explores a selection of historical and more contemporary examples of organizing around social reproduction over the last 40 years, the period of neoliberal or financialized capitalism. These examples anchor the concrete definition of social reproduction developed in section two which highlights the centrality of social reproduction politics to working class struggles more broadly. Informed by this analysis, the third section reflects on these struggles in the context of recent and widespread organizing and mobilizing around the intensifying global crisis of care/social reproduction, and the relevance of these struggles and this analysis in the context of the global coronavirus pandemic and protests to defend Black lives; two intimately related issues which, at the time of writing, have brought these contradictions into even starker relief. The fourth and concluding section reflects on these struggles, and the possibilities inherent within them for alternatives to the contemporary regime of social reproduction beyond the confines of capitalist social relations.

Austerity, social reproduction, and social movements

Neoliberal restructuring

Alongside the assault on wages and working-class organization in all of its forms over the last half century neoliberal economic restructuring has also rolled back state entitlements around the world. In addition to the reduction, re-allocation (towards capital) and/or elimination of protections for and public investment in key national industries (agricultural extension services for small-scale farmers for example), related infrastructure, and workplace protections, this roll back included reductions in public spending for the provision of public services like water and sanitation, housing, healthcare and education, and their commodification, and partial or outright privatization. These latter reforms effectively instituted a new regime of capitalist social reproduction, argues Nancy Fraser (2017), distinct from those of late nineteenth century liberal competitive capitalism, which largely left workers to fend for themselves, and twentieth century state-managed or welfare state capitalism, which saw an increased role for government in ensuring the reproduction of labour.

While countries in the North have experienced a roll-back of the welfare state which reached its pinnacle in the post-war period, the most painful experiences of restructuring have undoubtedly been in the South where entitlements were thin or non-existent even prior to the institution of neoliberal reforms. Since the late 1970s, Southern states have been subject to harsh economic restructuring in various guises, Structural Adjustment Programs (SAPs) being the most widely known and frequently critiqued. These programs were originally instituted in the interests of maintaining debt repayment to largely Northern creditors following three shocks: the debt crisis precipitated by profligate lending to Southern states in the 1960s and 1970s, the oil crisis in the latter decade, and the massive interest rate hikes of the early 1980s. Following the neoliberal policy prescriptions known as the "Washington Consensus" (Williamson 2004) and the only superficially distinct variants which came in its wake (Peck and Tickell 2002), successive rounds of restructuring have systematically undermined both production and social reproduction in debtor countries.

Social movements against austerity

Across the South these reforms gave rise to what Walton and Seddon (1994, p. 39) call *austerity protests*: "large-scale collective actions including political demonstrations, general strikes, and riots, which are animated by grievances over state policies of economic liberalization implemented in response to the debt crisis and market reforms urged by international agencies." From 1976 to 1992, with a peak between 1983 and 1985, there were at least 146 of these "IMF Riots" around the world—so-called given their link to the leading International Financial Institutions (IFI), the International Monetary Fund (IMF), and World Bank, that play powerful roles in international financial stability and development policy respectively (Walton and Seddon 1994, p. 42).

As Walton and Seddon (1994) point out, the motivation for these protests were both the actual distributional impacts of SAPs and the perception of injustice, and imperialist interference which accompanied them. Although unevenly experienced, these policies have had widespread governance implications, and have favoured particular sections of domestic economic and political elites in

Southern states who have similarly become targets of protest. Meanwhile, the negative impacts of increases to the cost of food, water, and other basic necessities with the elimination of price controls, subsidies, and other protections, and the commodification of public services have been far reaching, disproportionately borne by the poor majority, women especially.

Much like in previous eras of transition and change, and although elements of the contemporary regime of social reproduction have become normalized in some respects, there has been sustained and intense opposition to the remarkably consistent program of neoliberal economic restructuring—despite its constantly changing face—which has characterized the past 40 years of global development policy. Concentrated in Latin America, a second wave of these austerity protests had some success around the turn of the millennium alongside the anti-globalization movement mobilizations in both the North and South around summits of the G20, and the World Trade Organization, and global days of action in opposition to the IMF and World Bank (Wood 2013). For example, in 2000 a sharp increase in tariffs following water privatization in the Bolivian capital Cochabamba sparked widespread protest leading to policy reversal and the strengthening of progressive forces in the country which would contribute to the election of the Indigenous-led "Movement Towards Socialism" later in the decade (Spronk 2017; Webber 2011).

Former World Bank Chief Economist Joseph Stiglitz, who parted ways with the Bank in 1999 over his opposition to its policies, outlined the IMF/World Bank Four Step austerity program in an interview with the Guardian newspaper in 2001. After steps one and two, privatization and capital market liberalization,

> The IMF drags the gasping nation to Step Three: market-based pricing—a fancy term for raising prices on food, water, and cooking gas. This leads, predictably, to Step-Three-and-a-Half: what Stiglitz calls "the IMF riot."
>
> The IMF riot is painfully predictable. When a nation is, "down and out, [the IMF] squeezes the last drop of blood out of them. They turn up the heat until, finally, the whole cauldron blows up,"...
>
> You'd almost believe the riot was expected. And it is... [in] several documents from inside the World Bank. In one... the Bank several times suggests—with cold accuracy— that the plans could be expected to spark "social unrest" (Palast 2001 n.p.)

This all before Step Four as markets in the South are opened up to free trade and competition with the world economy with well-established negative effects.

In the context of the triple crises of finance, food, and fuel that followed the 2007–2008 financial crisis (Murphy and Wise 2013), Southern economies, now opened up to global markets, were left highly vulnerable as a renewed global austerity push precipitated another round of outrage. This time the North was not spared as protests, particularly around food, erupted in countries around the world from "pasta protests" in Italy to "tortilla protests" in

Mexico, and others in West Africa, Eastern Europe, and the Philippines (Patel 2009; Wood 2013).

In the decade after the Great Recession, several waves of protest took place around an array of interconnected issues as austerity and financialization delivered little by way of economic growth in the real economy (Tooze 2018) and reactionary, right-wing justifications for rising poverty and inequality gained ground. This round of protests included Occupy; #Black Lives Matter; #RhodesMustFall and #FeesMustFall; the struggles of water and land protectors at Standing Rock, the Wet'suwet'en, and the broader movement for decolonization; the worldwide feminist strikes (which it has been argued represent a true third feminist wave by Arruzza (2018; see also Davis et al. 2017) the climate strike movement; and protected and unprotected strikes by teachers, nurses, and other care workers as well as workers in the gig economy, and other precarious jobs the world over.

Even more recently the massive protests that have rippled across the South and North contesting the broad organization of capitalist society have integrated if not centered social reproduction demands for basic necessities and more. Among these we can count the millions of people in the streets from India to Chile, Lebanon, Iraq, Iran, Ecuador, Bolivia, Sudan, Hong Kong, and elsewhere, demanding a more democratic—not just in terms of civil and political rights— and equitable world as their ability to survive and thrive continues to be undermined in the interests of sustaining capital accumulation (McNally 2020).

Social reproduction: Definitions and theory

Social Reproduction Theory argues that the systematic undermining of social reproduction just described is both inherent in capitalism, and essential to its perpetuation. As Nancy Fraser (2017, p. 21) argues, the crisis of care or social reproduction "refers to the pressures from several directions that are currently squeezing a key set of social capacities: the capacities available for birthing and raising children, caring for friends and family members, maintaining households, and broader communities, and sustaining connections more generally." This gendered, affective and material labour of social reproduction is, she argues "indispensable to society. Without it there could be no culture, no economy, no political organization. No society that systematically undermines social reproduction can endure for long" (Fraser 2017, p. 21). The undermining of social reproduction is thus a key constitutive element of contemporary crises.

Historically, the undermining and/or destabilization of social reproduction have led to periods of crisis marked by intense protest. Commenting on the various waves of austerity protests discussed above, Patel and McMichael (2014, p. 12) argue that the spike in the price of food and water as a result of liberalization and the transition to market pricing represents an attack on the norms of what historian E.P. Thompson (1971), commenting on the transition to capitalism in England, called the "moral economy:" "the cluster of political and pre-political ideas circulating within society that governed the natural and desirable means of the distribution of common wealth." Similar to arguments made by liberal development economist and scholar of famine Amartya Sen (1981), Patel and McMichael argue that the inaccessibility of food and services resulting from often abrupt price increases in subsequent transitions to capitalism or moments of crisis within it can also be understood as "entitlement failure" around which struggles ensue leading to the establishment of new norms or social agreements, however inequitable. In reflecting on Thompson and Sen's work, among others, Patel and McMichael (2014, p. 13) point out that historically "the replacement of one set of entitlements with another was not smooth or swift, but fragmentary, disjointed, and sometimes violent." Indeed, protest often becomes the political tactic of choice when other avenues of political engagement are absent or appear ineffective.

Similarly, social reproduction scholar Nancy Fraser refers to these contests as "boundary struggles" which characterize the class antagonism at the heart of the crisis of social reproduction and of capitalism itself. At the root of this crisis is capital's desire to constantly diminish the price it pays for labour as the essential input in the capitalist production of value; and indeed it strives to pay a price below the cost of its replacement or avoid paying for it altogether (Fraser is careful to point out that capital has a similar attitude towards non-human nature and the inputs it is able to expropriate from it). As a result, the gendered labour of social reproduction, whether unpaid or paid, in the home, community, and/ or social institutions, and the bodies that perform it are constantly devalued and the means by which it is carried out undermined. With the assault on the working class that began in the 1970s the resources available to support the renewal of the labouring classes were hollowed out by capital's drive for accumulation, increasingly free of its social obligations as a result of widespread cuts to sectors essential for social reproduction. In such cases, Fraser 2017, p. 24) argues, "the logic of economic production overrides that of social reproduction, destabilizing the very social processes on which capital depends—compromising the social capacities, both domestic and public, that are needed to sustain accumulation over the long term. Destroying its own conditions of possibility, capital's accumulation dynamic effectively eats its own tail."

Taking the production of the commodity of labour power, and specifically working-class experience, as its starting point, SRT helps to rectify the overly narrow focus on relations of production—i.e. the workplace—that dominate many understandings of capitalism and social struggle which privilege class, and—somewhat ironically—those that deny the utility of the category altogether. In integrating production and reproduction, SRT conceives of class struggle as unfolding not just in the realm of the former—the workplace, the factory—but also within the realm of the latter—the broader social totality of capitalism constituted by all variety of social relations (Ferguson 2016).

This expansive conception of capitalism is not an attempt to privilege struggles in the home or community over those in the workplace but to situate all struggles for life-making within the historically constituted, messy totality of capitalist social relations. This is also not to privilege struggles around exploitation versus those around oppression but rather to insist on an integrative analysis of the two, given that the latter have been and continue to be essential to the reproduction of a differentiated labour force—into which relations of race, gender, and nationality/citizenship figure heavily—in which some workers' labour comes at a lower cost than others. This is evident, for example, in the maintenance of a highly restrictive and exploitative system of global migration alongside the relative freedom enjoyed by capital (Federici and Jones 2020; McNally and Ferguson 2015). Social relations of oppression are thus upheld systemically within capitalism in part due to their role in the development of variously segmented labour markets which are particularly efficacious in facilitating the extraction of surplus value from human labour.

Developing an analysis of a capitalist totality that integrates both production and reproduction is not just a descriptive or analytical academic exercise, however. As Meg Luxton (2006, pp. 36–37) argues, it also "allows for an explanation of the structures, relationships, and dynamics that produce those activities" which is vital to the process of developing tools to contest them. Indeed, SRT's integrative theorization of production and reproduction within a capitalist totality offers much to understanding both historical waves of protest against austerity and those unfolding in the present, and to the task of building working class unity to fight back against what is already shaping up to be a renewed austerity project in the aftermath of the coronavirus pandemic, and the continued devaluation of racialized bodies in both various national contexts and globally.

Crises of care, COVID-19, and the Carceral State

As these words are put to page the world remains in the throes of the coronavirus pandemic amidst renewed calls for increased social spending in light of the pandemic and for social and economic justice more broadly. The pandemic has intensified the ongoing crisis of care/social reproduction in late capitalism and lays bare, perhaps for the first time for some, the contradiction in capitalism between profit and "life-making," (Bhattacharya 2020b). At the heart of this contradiction is that while capital depends on workers to produce commodities (hence the present calls for people to get back to work), as Tithi Bhattacharya (2020a) points out it is a reluctant dependence. On the one hand capital must ensure workers' biological and social reproduction not only tomorrow but into the future as well, but on the other the profit motive necessitates that capital constantly push in the opposite direction such that all the paid and unpaid labour of reproducing workers daily and generationally is devalued and taxes kept low so that the institutions that support these activities remain underfunded, and/or under threat of partial or outright privatization. The reluctant dependence of this contradiction is summed up well by Wendy Brown (2015, pp. 104–105) who argues that the paid and unpaid, affective and material labour of social reproduction—mostly performed by (often racialized) women—acts as "the unavowed glue for a world whose governing principle cannot hold it together." COVID-19 has only brought the capitalist crisis of care and social reproduction into sharper relief, both in the pandemic's effects even more clearly exposing how capitalism differentiallwy values some labouring bodies over others and as state responses to the pandemic disproportionately benefit capital.

Pandemic, crisis, and social reproduction

Some might argue that the crisis caused by the COVID-19 pandemic is different, that it is caused by a virus that does not discriminate between human beings, and that we are all in this together. This virus and the disease it causes start in the biological realm to be sure, but as Rob Wallace (2016) and Mike Davis (2006; 2020) have pointed out, the recent increase of health threats from the human-animal-ecosystems interface (HAEI) in the form of zoonotic diseases (zoonoses) cannot be understood without considering contemporary capitalism and indus-trial agriculture, and land use change in particular. Thus, in both its origins and in its transmission, as Kim Moody's (2020) work on global supply chains has shown, COVID-19 is tightly bound up with the social relations through which contemporary capitalism organizes the production and reproduction of life on planet earth. But, as SRT makes vividly clear, it is in the virus' impact, and our response to it that it is most apparently and deeply integrated with the inequitable social relations by which our political economy is organized, none the more so than in the realm of social reproduction.

There is indeed no end to the examples we can refer to which demonstrate the uneven impact of the pandemic and the distribution of its effects according to pre-existing, and now amplified inequities or those newly emergent. We only need to look at the statistics on those most affected by the virus which, even in a relatively equitable country like Canada, are heavily skewed towards working class people of colour and, though the virus seems to have more serious effects on men once they are infected, women make up a greater share of cases given their over representation in care and other essential work (Fraser 2017). When we look elsewhere, to countries in the North with higher inequality, and to the South the picture is even more bleak. As diametrically opposed organizations as Oxfam (Jacobs and Lawson 2020) and the World Bank (Lakner et al. 2020) both argue—with important differences in their analysis, conclusion and policy prescriptions of course—that the impact of COVID-19 on poverty and inequality in the South will be devastating. For example, in the face of years of the hollowing out of the public provision of water, sanitation, and health, in the absence of adequate or any savings at all amidst stagnating wage growth, massive unemployment, and/or precarious work worsened by the pandemic, and skyrocketing household debt, Oxfam (Jacobs and Lawson 2020) estimates that half a billion people risk falling into poverty in developing countries alone (See Sumner, Ortiz-Juarez and Hoy 2020 for further analysis of the impact of COVID-19 on poverty and inequality).

Who gets protected and how?

Despite the impacts of COVID-19 thus far and the dire warnings of what is to come, the implementation of welfare measures in response to the pandemic has been remarkably uneven across the world. Even in the best of cases, states have largely directed their responses to the needs of capital as they socialize the costs of any measures taken. Like in the aftermath of the Great Recession, where "market protection" favoured capital over labour, providing guarantees to lenders rather than to needy workers in search of housing and livelihoods for example (Soederberg 2016), there is already evidence that the response to the coronavirus is following much the same formula. In the UK the private sector is being brought in to fill in for the NHS (Garside and Neate 2020). Here in Canada, in the province of Manitoba public sector cuts earlier in the pandemic were wide and deep (Hajer et al. 2020) while the shock therapy just announced in Alberta has lowered the corporate tax rate far below the Canadian average, slashed spending, and clamped down on democratic dissent while privileging the rights of capital, fossil capital in particular (Anderson 2020). Similarly, in Ontario the conservative government is seeking to roll back, and/or suspend workers' rights ostensibly in the interests of supporting public health measures (Khurram 2020). South of the border in the United States, as Robert Brenner (2020, p. 20) notes, in their truly unprecedented bailout of American capital since the pandemic began, the Federal Reserve and Treasury Department "'have essentially been socializing credit risk', and in the process have been 'creating a new moral hazard'" in the expectation that the state will facilitate market function and access to credit irrespective of corporate balance sheets. In this Brenner argues that we are witnessing a repurposing of the state in the interests of capital yet again, just as we did in the crisis in the 1970s and after the Great Recession, this time to a financialized Keynesianism which has allowed billionaires to rapidly increase their wealth in an obscene fashion while on main street lines for food banks are miles long. Even when less overt, like in South Africa, which relative to the size of its economy has one of the largest stimulus responses in the world, the response is disproportionately skewed towards capital while the country's underlying market orthodoxy, which has long informed its approach to social protection, undermines the potential of its seemingly impressive attempts to help its citizens and transform its economy (C19 People's Coalition 2020).

All this is not to overlook the unprecedented response on the part of many states to the pandemic, however. There have indeed been some unprecedented interventions, particularly those in the South which include Vietnam, Laos, Cuba, and the Indian State of Kerala (Tricontinental 2020), and what seemed impossible yesterday now seems necessary. But, as Gareth Dale and Tithi Bhattacharya (2020) argue, "If in liberal normality the welfare and repressive spheres [of the state], although connected, are most often experienced as separate, right now they're scrambled together in unprecedented ways." While there have been notable trends in this direction over the past decade as global inequality continues to rise and reactionary politics threaten democratic institutions, in addition to the market disciplinary measures discussed above, which many argue are the more commonly used techniques of governmentality in our age of market individualism (Best 2013), the coercive apparatus of the state has only continued to grow and tighten in the response to COVID-19.

This is perhaps felt most acutely in the South, even in the ostensibly liberal democracies. For millions of people in the South in places like Kenya, South Africa, and India, among others, livelihoods have been destroyed, access to basic necessities have virtually disappeared and, early on, deaths at the hands of the police equalled or outnumbered those from COVID-19 while brutality complaints number in the hundreds and thousands. For example, in South Africa the recent and long-called for increases in the magnitude of investment in health systems and social grants (cash transfers) to individuals and households, among other interventions made in response to COVID-19, were accompanied by the addition of some 78 000 SADF soldiers to the pandemic response. While government investments were welcomed, the widespread consensus is that they are too little, and the addition of military personnel considered wasteful and tone deaf; between March 26th, when the country's initial lockdown came into place, and the end of April, the military and police killed at least eight people, and more than 200 cases of police brutality were recorded (Shoki 2020). In the early stages of lockdown enforcement, as in Kenya, deaths by police outnumbered those by the virus.

But let us not forget how the response to the pandemic has played out in the North in this period either: the long overdue preliminary data from Canada's largest city, Toronto, clearly shows the disproportionate impact of the virus and disciplinary measures including fines and arrests on people of colour, the poor, and those made most vulnerable in our societies (Amin and Bond 2020; Bouka and Bouka 2020; City of Toronto 2020). Even more problematically, as this data shows, the balance of governmental techniques experienced by a population appears to be closely related to their location and composition: women, racialized people, and others made precarious in capitalist society have long and disproportionately felt the simultaneous dread of water and electricity cut-offs and insecure housing as well as the sting of the baton and the grief of a lost child cut down by police. At their very worst, we have witnessed the intensification of past and present inequities as Black lives are increasingly snuffed out by state violence. As a result, the words "Black Lives Matter" once again reverberate in cities not only in the United States but around the world as thousands take to the streets to protest police violence against Black and people of colour, and its intensification under the pandemic lockdown (Leicester and Jordans 2020; Taylor 2020).

There are significant tensions, Dale and Bhattacharya (2020) point out— but also alarming parallels in terms of impact—between these heavy-handed approaches to state assistance and the abdication, like that of Donald Trump, Narendra Modi or Jair Bolsonaro, of mandating appropriate public health measures and income supports, themselves disastrous moves especially for the oppressed and exploited, and a move to relax public health measures in the interests of re-opening economies. While these may be a few of the more egregious examples, they are not the only places where coercive and punitive force alongside assistance measures, and/or a too-rapid return to normalcy for the sake of the economy have proceeded or been considered. Countries like Canada have also experienced this creeping enforcement overreach in the context of millions falling through the cracks of welfare interventions. And as time wears on, pressure mounts, and steps are taken to re-open the economy, the burden of which will, as it already has, continue to be borne everywhere by those suffering gender, race, and status discrimination.

Conclusion: Coronavirus, race, and gender

As outlined at the beginning of this chapter, even before the COVID-19 crisis we were witnessing an uptick in organizing around the crisis of social reproduction and care: the "third feminist wave" of feminist strikes, climate strikes, BLM and BIPOC movements for racial justice, decolonization, care worker strikes, massive protests in the South (and North) around a whole range of issues but which centrally integrate demands for access to the means of social reproduction and more. These dynamics all point to the existence and development of an organizational infrastructure, and set of resources which seeks to push back against the already underway coronavirus response from above to introduce a new wave of austerity, and put pressure on working class life while prioritizing capital in general and finance capital in particular. In other words, the intensification of organizing around social reproduction and care increasingly appears to be at the centre of struggles for human survival in the short term and a key site of struggle which has the potential to contribute to limiting how much capital can get away with while the working-class continues to fight for and propose alternatives to yet another repackaging of the status quo.

While much appears to be business as usual amidst the pandemic what is different about the present moment is that the coronavirus is shining a light, if a tragic one, on the relationship between capitalism and care/social reproduction. And despite capitalism's temporary prioritization of life-making in the context of the pandemic—a focus that humanity urgently needs more generally—it is more concerned with saving the economy (Bhattacharya 2020a). Without question, however, and hopefully, we are seeing a marked intensification in contemporary organizing around social reproduction. We can think of everything from care-mongering and other mutual aid groups—with varying degrees of political critique—to "illegal" yet absolutely necessary, water and utility reconnections, rent strikes, PPE drives, drive-by and in-person protests at legislatures, hospitals, farms, factories, warehouses and prisons, wildcat strikes by organized and unorganized workers, strikes targeting the repurposing of plant and equipment towards the public health response, court cases on the part of unions around worker rights and protections, and, perhaps most notably, the protests against the historical and ongoing abuse of racialized people by police and their intensification during the pandemic.

In all of these efforts it is apparent that organizing, protest, and engagement in policy processes in the present builds upon dense networks of organizing and knowledge creation, and sharing that extend far into the past. Contemporary efforts build upon these pre-existing relations of community care and organization in an effort to emerge from the pandemic with more demands for collective care, and to fight austerity in the context of the economic crisis which is already here and will continue to intensify. Put another way, many of these movements appear to emphasize nurturing and defending the wellsprings of social reproduction from below which have risen to the surface or sprung up anew in the midst of this pandemic. It is likewise important that we render these experiences critical in order to further develop and cement the bonds of social solidarity, and forms of organization necessary to hold the state and capital at bay, and prioritize human life such that we can emerge from this multi-pronged crisis stronger than before.

In this way SRT's analysis of struggles around social reproduction is not only useful because it helps us to understand the differentiated reproduction of labour power under capitalism. Rather, in providing us with some important analytical tools to fight back against capital it also theorizes what socially reproductive labour is and can be outside of and beyond capitalism. In Tithi Bhattacharya's (2020b) most recent conceptualization, "the best way to define social reproduction is the activities and institutions that are required for making life, maintaining life, and generationally replacing life," what she calls "'life-making' activities," a term which has been used throughout this chapter. While maintaining but also helping us to better understand the importance of struggles over wages in the workplace, SRT theorizes these struggles in relation to those around social reproduction within a broader capitalist totality. As Bhattacharya has elaborated elsewhere, the wages people work for under capitalism are not ends in themselves but rather are used to purchase the means of social reproduction, of life-making, to provide the worker and her family with the best possible life. In re-emphasizing life-making as such the struggle against capitalism, and capital in general is re-cast, and the possibilities for not just the alternative organization of social reproduction but of capitalism as a whole are renewed and expanded, an imagination which is sorely needed as the pandemic rages on.

References

Amin, F. and Bond, M. (2020). COVID-19 disproportionally impacting Black communities, people of colour in Toronto. *City News* [online], 9 July 2020. Available from: https://toronto.citynews.ca/2020/07/09/race-data-covid-toronto/ [accessed 11 July 2020].

Anderson, D. (2020). Alberta bets on infrastructure spending, corporate tax cuts to spur recovery | *CBC News* [online], 29 June 2020. Available from: https://www.cbc.ca/news/canada/calgary/kenney-economic-reboot-announcement-1.5631088 [accessed 9 July 2020].

Arruzza, C. (2018). From Women's strikes to a New class movement: The Third feminist wave. *Viewpoint Magazine.*

Bakker, K.J. (2010). *Privatizing Water: Governance Failure and the world's Urban Water Crisis.* Ithaca, N.Y: Cornell University Press.

Best, J. (2013). Redefining poverty as risk and vulnerability: Shifting strategies of liberal economic governance. *Third World Quarterly*, 34(1), pp. 109–129.

Bhattacharya, T. (2020b). Social reproduction and the pandemic, with tithi bhattacharya. *Dissent Magazine* [online]. Available from: https://www.dissentmagazine.org/online_ articles/social-reproduction-and-the-pandemic-with-tithi-bhattacharya [accessed 10 June 2020].

Bhattacharya, T. (2020a). Social reproduction theory and why we need it to make sense of the COVID-19 crisis. *URPE* [online]. Available from: https://urpe.org/2020/04/02/ social-reproduction-theory-and-why-we-need-it-to-make-sense-of-the-corona- virus-crisis/ [accessed 22 June 2020].

Bouka, A-A. and Bouka, Y. (2020). Canada's COVID-19 blind spots on race, immi- gration and labour. *Policy Options* [online]. Available at: https://policyoptions.irpp. org/magazines/may-2020/canadas-covid-19-blind-spots-on-race-immigration-and- labour/ [accessed 22 June 2020].

Brenner, R. (2020). Escalating plunder. *New Left Review*, 123, pp. 5–22.

Brown, W. (2015). *Undoing the Demos: Neoliberalism's Stealth Revolution.* MIT Press.

C19 People's Coalition. (2020). Media alert: C19 People's Coalition online press con- ference. *C19 People's Coalition* [online]. Available from: https://c19peoplescoalition. org.za/media-alert-c19-peoples-coalition-online-press-conference/ [accessed 7 May 2020].

City of Toronto. (2020). Toronto Public Health releases new socio-demographic COVID-19 data. *City of Toronto* [online]. Available from: https://www.toronto. ca/news/toronto-public-health-releases-new-socio-demographic-covid-19-data/ [accessed 1 August 2020].

Dale, G. and Bhattacharya, T. (2020). Covid capitalism. *Spectre Journal* [online], 2020. Available from: https://spectrejournal.com/covid-capitalism/ [accessed 5 May 2020].

Davis, A. et al. (2017). Beyond lean-in: For a feminism of the 99% and a militant interna- tional strike on March 8. *Viewpoint Magazine* [online]. Available from: https://www. viewpointmag.com/2017/02/03/beyond-lean-in-for-a-feminism-of-the-99-and-a- militant-international-strike-on-march-8/ [accessed 20 April 2020].

Davis, M. (2020). The Monster Enters. *New Left Review*, 122, pp. 7–14.

Davis, M. (2006). *The Monster at our Door: The Global Threat of Avian Flu.* New York: Macmillan.

Federici, S. and Jones, C. (2020). Counterplanning in the crisis of social reproduction. *South Atlantic Quarterly*, 119(1), pp. 153–165.

Ferguson, S. (2016). Intersectionality and social-reproduction feminisms. *Historical Materialism*, 24(2), pp. 38–60.

Fraser, N. (2017). Crisis of care? On the social reproductive contradictions of contem- porary capitalism. In: Tithi Bhattacharya, ed. *Social Reproduction Theory: Remapping Class, Recentering Oppression.* London: Pluto Press, pp. 21–36.

Garside, J. and Neate, R. (2020). UK government 'using pandemic to transfer NHS duties to private sector.' *Guardian* [online]. 4 May 2020. Available from: https://www. theguardian.com/business/2020/may/04/uk-government-using-crisis-to-transfer- nhs-duties-to-private-sector [accessed 10 June 2020].

Hajer, J and Fern, L. (2020) *Austerity and COVID-19: Manitoba government creating, not solv- ing, problems | CBC News* [online]. 21 April 2020. Available from: https://www.cbc.

ca/news/canada/manitoba/manitoba-government-economy-covid-19-1.5539666 [accessed 27 April 2020].

Jacobs, D. and Lawson, M. (2020). Dignity not destitution: An 'Economic rescue plan for all' to tackle the coronavirus crisis and rebuild a more equal world. *Oxfam*. London: Oxfam.

Khurram, S. (2020). Delayed, negligent, and ineffectual: Doug Ford's botched response to the COVID-19 pandemic. *Canadian Dimension* [online]. Available from: https://canadiandimension.com/articles/view/doug-fords-botched-response-to-covid19 [accessed 6 August 2020].

Kishimoto, S and Petitjean, O. (2017). *Reclaiming Public Services: How Cities and Citizens Are Turning Back Privatization*. Amsterdam: TNI.

Lakner, C., Mahler, D.G., Negre, M and Beer, E. How much does reducing inequality matter for global poverty? *Global Poverty Monitoring Technical Note*. 13. Washington: World Bank.

Leicester, J. and Jordans, F. A look at Black lives matter protests from around the world. *Globe and Mail* [online]. 6 June 2020. Available from: https://www.theglobeandmail.com/world/article-a-look-at-black-lives-matter-protests-from-around-the-world/ [accessed 8 June 2020].

Luxton, M. (2006). Feminist political economy in Canada and the politics of social reproduction. In: Luxton, M. and Bezanson, Kate, eds. *Social Reproduction: Feminist Political Economy Challenges Neo-Liberalism*. McGill-Queen's Press - MQUP, pp.11–44.

Luxton, M. and Bezanson, K. (2006). *Social Reproduction: Feminist Political Economy Challenges Neo-Liberalism*. McGill-Queen's Press - MQUP.

McDonald, D.A. (2016). *Making Public in a Privatized World: The Struggle for Essential Services*. Zed Books Ltd.

McDonald, D.A., ed. (2015). *Rethinking Corporatization and Public Services in the Global South*. London: Zed.

McNally, David and Ferguson, Sue. (2015). Social reproduction beyond intersectionality: An interview. *Viewpoint Magazine* [online]. Available from: https://www.viewpointmag.com/2015/10/31/social-reproduction-beyond-intersectionality-an-interview-with-sue-ferguson-and-david-mcnally/ [accessed 7 May 2020].

McNally, D. (2020). The return of the mass strike. *Spectre Journal* [online]. Available from: https://spectrejournal.com/the-return-of-the-mass-strike/ [accessed 10 June 2020].

Moody, K. (2020). How "Just-in-Time" Capitalism Spread COVID-19. *Spectre Journal* [online]. Available from: https://spectrejournal.com/how-just-in-time-capitalism-spread-covid-19/ [accessed 10 June 2020].

Murphy, S. and Wise, T.A. (2013). Resolving the food crisis : The need for decisive action. *Aljazeera* [online]. 30 January 2013. Available from: https://www.aljazeera.com/indepth/opinion/2013/01/201312915630857878.html [accessed 17 May 2020].

Nash, J. (1994). Global integration and subsistence insecurity. *American Anthropologist*, 96(1), pp. 7–30.

Palast, G. (2001). IMF's four steps to damnation. *The Observer* [online], 29 April 2001. Available from: https://www.theguardian.com/business/2001/apr/29/business.mbas [accessed 17 June 2020].

Patel, R. (2009). Food Riots. In: Immanuel Ness, Dario Azzellini, Marcelline Block, Jesse Cohn, Clifford D. Conner, Rowena Griem, Paul LeBlanc, Amy Linch, eds. *The International Encyclopedia of Revolution and Protest 1500 to the Present*. N.p.

Patel, R. and McMichael, P. (2014). A political economy of the food riot. In: Pritchard, D. and Pakes, F., eds. *Riot, Unrest and Protest on the Global Stage*. London: Palgrave Macmillan UK, pp. 237–261.

Peck, J. and Tickell, A. (2002). Neoliberalizing space. *Antipode*, 34(3), pp. 380–404.

Sen, A. (1981). *Poverty and Famines: An Essay on Entitlement and Deprivation.* Oxford University Press.

Shoki, William (2020). The class character of police violence in South Africa. *Verso Blog* [online]. Available from: https://www.versobooks.com/blogs/4795-the-class-character-of-police-violence-in-south-africa [accessed 26 July 2020].

Soederberg, S. (2016). Introduction—Risk management in global capitalism. In: S. Soederberg (ed.), *Research in Political Economy.* Bingley, U.K: Emerald Group Publishing Limited.

Spronk, S. (2017). Class struggle and resistance in latin America. In: Veltmeyer, H. and Bowles, P., eds. *The Essential Guide to Critical Development Studies.* New York, NY: Routledge, pp. 279–287.

Sumner, A., Ortiz-Juarez, E. and Hoy, C. and UNU-WIDER. (2020). *Precarity and the pandemic: COVID-19 and poverty incidence, intensity, and severity in developing countries* [online]. 77th ed. WIDER Working Paper. UNU-WIDER.

Taylor, K.-Y. How do we change America? *The New Yorker* [online]. 8 June 2020. Available from: https://www.newyorker.com/news/our-columnists/how-do-we-change-america [accessed 10 June 2020].

Thompson, E.P. (1971). The Moral Economy of The English Crowd In The Eighteenth Century. *Past and Present*, 50(1), pp. 76–136.

Tooze, A. (2018). *Crashed: How a Decade of Financial Crises Changed the World.* New York: Penguin.

Tricontinental, I. (2020). *Coronashock and Socialism* [online]. New Delhi: Tricontinental Institute.

UNICEF and WHO. (2019). *Progress on Household Drinking Water, Sanitation and Hygiene, 2000–2017.* New York: Unicef.

Wallace, R. 2016. *Big Farms Make Big Flu: Dispatches on Influenze, Agribusiness, and the Nature of Science.* New York: NYU Press.

Walton, J and Seddon, D. (1994). *Free Markets & Food Riots: the Politics of Global Adjustment.* Cambridge, Mass: Blackwell.

Webber, J.R. (2011). *Red October: Left-Indigenous Struggles in Modern Bolivia.* Boston: Brill.

Williamson, J. (2004). A short history of the Washington consensus. In: *From the Washington Consensus Towards a New Global Governance.* Barcelona: Fundación CIDOB.

Wood, L.J. (2013). Anti-world Bank and IMF riots. In: Snow, D. A. et al., eds. *The Wiley-Blackwell Encyclopedia of Social and Political Movements.* Oxford, UK: Blackwell Publishing.

World Bank. (2015). *FROM BILLIONS TO TRILLIONS: MDB Contributions to Financing for Development.* Washington: World Bank.

World Bank. (2003). *World Development Report 2004: Making Services Work for Poor People.* Washington: World Bank.

4

GLOBALIZATION AND THE TWIN SCOURGES OF ILLIBERALISM AND INEQUALITY

Syed Mansoob Murshed

Introduction

At the beginning of the new millennium, development studies almost single-mindedly focused on the objective of poverty reduction. There were those who emphasized good governance, and democratic values[1]. This coincided with the so-called 'third wave' of democratization following the end of the cold war (Huntington 1993), and gradually more and more nations across the planet became wedded to the idea of multi-party elections. Others in development studies were, as ever, more concerned with enhancing material well-being, the chief vehicle for doing so being economic growth. It was believed that economic growth constituted the principal avenue via which sustainable poverty reduction can be attained in low-income developing countries. Redistribution without enlarging the cake only served to make the already poor more equal. Thus, growth is a necessary condition for poverty reduction in low-income countries. Growth would always diminish poverty as long as some of the benefits of growth trickle down to the poor, even if its principal beneficiaries are the wealthy.

Truly pro-poor growth, however, such as in the notion advocated by Kakwani and Pernia (2000) suggested that growth to be truly pro-poor it must disproportionately benefit the poorer segments of society; thus requiring an improvement in the distribution of income.[2] In the past two decades there has been considerable success in reducing poverty in the developing world as economic growth rates gathered pace, but this has been at the expense of rising inequality; see Jayadev, Lahoti and Reddy (2015). In the global North, inequality has also risen considerably, along with the rise in precarious employment (Standing 2011). Furthermore, the last two decades have marked an era of prolonged real wage compression in developed economies, particularly since the financial crisis of 2008. The combined effects of wage compression and precarious employment (particularly in the two large English speaking nations, the USA and the UK) has meant that the spectre of poverty has once again returned to haunt Western economies, and non-contractual precarious employment or self-employment has transformed vast swathes of the labour force into something akin to the daily wage labour in the global South.

A degree of economic inequality is inevitable in any society, reflecting the price paid for the incentivization of risk and work effort, but inequality which perpetuates discriminatory practices against certain groups in society and prevents the under-privileged from rising by the fruits of their own effort is described as *inequality of opportunity*. This form of inequality is both undesirable and economically inefficient (Stiglitz 2012; Roemer 1998). The rise in inequality is a globally ubiquitous phenomenon, as is the growing tide of precarious work; attention needs to be paid to these phenomena, shifting our gaze away from a sole focus on poverty reduction. The next section examines these trends in the context of globalization.

This brings us to developments on the democratic front which ultimately governs the environment within which policy is formulated and implemented. Thus, economics and politics are inseparable. Multi-party electoral competition is now an almost universal periodic ritual, albeit subject to manipulation and violence in many cases. The fall of the Berlin Wall in 1989 created expectations, for a long wave of liberalism, but the past decade has seen the rise of populism and authoritarianism, which increasingly flout the *liberal* principles of a liberal-democracy, despite permitting regular elections. It is the liberal aspect of liberal-democracy that has been undermined. There appears to be a global trend towards illiberal democracies (Zakariah 1997), and this tendency also requires our attention. Indeed, as Rodrik (2017) indicates, there is no good reason to expect democracy and liberalism to necessarily cohabit, and in our globalized world they seem to be increasingly becoming strange bedfellows. I consider these developments in the third section of this chapter. The important point to bear in mind is that the twin phenomena considered in the chapter are inter-related, as there is a causal link running between the rising inequality and the surge in support for illiberal populism. The final section of the chapter offers a synthesis and concludes with a cautionary note directed against viewing de-globalization as a universal panacea for populism and inequality.

Globalization and inequality

Economic globalization refers to the intensity of international trade in goods and services, as well global financial flows. Increases in international trade nearly always have distributional consequences. After an expansion of trade, the factors of production employed in the exportable sector will witness a rise in their relative remuneration. This is because the exportable sectors of the economy expand, and the import-competing sectors contract, after increased international trade. Hence, greater globalization produces winners and losers.[3] For example, with the end of the multi-fibre agreement which governed and restricted the import of garments from developing countries into OECD nations, the remaining European Union producers (in countries like Portugal) were hurt by Chinese goods, so they successfully lobbied for, and obtained a re-imposition of import controls. This raises the question as to whether society is wishes to compensate the losers from increased trade. Traditionally it was believed that if there was economic growth following enhanced trade, the gainers gain is greater than the loss of the losers; there is a potential for compensating the losers, provided enough political will exists to affect the re-distribution.

Financial globalization and open capital markets are regarded to be less benign. They are believed to promote financial crises, particularly in developing countries. For example the recent crisis in the Southern European (Greece, Spain, and Portugal) and Ireland part of the euro-zone can be linked to capital mobility, a common lower interest rate leading to unsustainable financial flows and debt accumulation, which, once the financial boom was over, led to massive contractions in the economies of those countries. Of course, the faulty institutional architecture which fails to combine a supra-national currency with a sufficiently common fiscal policy also plays a crucial part.

One of the greatest challenges of our day is the growing global tide of inequality in income and wealth which has coincided with increasing globalization, and what needs to be done to redress this phenomenon. (Stiglitz 2012; Piketty 2014). Inequalities in income are more commonly measured, but inequalities in wealth are far greater, and the ownership of wealth is far more concentrated compared to income.[4] The chief concerns with these developments are to do with the income and wealth share of the richest 1% or 10% of the population. If we take a truly cosmopolitan view, treating the entire planet as a single entity, global inequality may have declined by about 2 percentage GINI[5] points between 1988 and 2008 to around 70.5 (Lakner and Milanovic 2015) as the two populous poor developing countries, China and India narrow their per-capita income gap with rich countries like the United States. The biggest beneficiaries of the change in the global distribution of income during this period were the middle classes in emerging economies like China and India; the greatest losers are the traditional working classes in the developed segments of the world. The authors point out, however, that inequality measures mask serious underestimation of the income of the top decile in the income distribution, who are often missed out in household surveys. Accompanying this rise in inequality is more precarious work, informalisation, and self-employment, which is more marked and noticeable in the global North, because precarious employment (and self-employment) had virtually become extinct in the developed segments of the world after the Second World War. In addition, the wealth share of the richest 1% in the world is greater than the rest of the population's (99%) total wealth according to some sources (OXFAM 2016).

The present world's super-rich are considerably richer in real terms than the super-rich of the past like the Carnegie's and Rockefellers (Goda 2014), implying a greater concentration of wealth at the top than ever before in human history. In the developing world, recent surges in economic growth have lowered poverty, propelling a good number of countries to middle income status, along with a doubling of people (about 3.2 billion) who are just above a decent global poverty line ($2.5 to $10 per-day in purchasing power parity dollars). But this vast swathe of the global citizenry occupy a precariously unstable zone, and risk descending back to poverty (Sumner 2016).

Dabla-Norris et al. (2015) and Ostry, Berg and Tsangarides (2014) show that the recent growth experiences of a cross-section of developed and developing countries suggest that inequality is actually harmful to growth prospects, contrary to the received wisdom that inequality (by permitting the rich to save and invest more in productive capital) oiled the wheels of growth. This could be because greater inequality leaves economies more prone to financial crises[6], greater inequality discourages human capital accumulation (education) among the poorer segments of society, and because inequality contains within it the seeds of conflict, which is harmful for growth. Moreover, the financial markets of today are less invested in productive manufacturing, and more in the casino capitalism of purely financial products, making the super-rich of today more like rentiers rather than the industrial capitalists of the past. On the causes of recent rises in income inequality, Dabla-Norris et al. (2015) point out three phenomena. First, unskilled labour saving technical progress which first decimated manufacturing jobs, and latterly in services as well; secondly, financial globalization (but not trade openness); thirdly, less regulation of labour markets, including the informalisation of work are the chief culprits. The nature of global manufacturing production has become increasingly fragmented, with components produced and shipped to different locations across the globe, turning nation states into regions.[7]

Financial globalization, and the greater mobility of capital contributes to greater inequality by lowering the bargaining power of labour under the threat of economic activities locating overseas (Furceri, Loungani and Ostry 2017). Rodrik (2018) also maintains that greater mobile capital shifts the burden of adjustment to economic shocks more to labour. Highly mobile international capital also has the effect of lowering corporate taxes via tax competition, and narrows the fiscal space and capacity of the state (see Rodrik 2018 and references therein). Technical progress has been cited as a major cause of job destruction, the displacement of production line workers, the automation of services, and the hollowing of the middle class by putting their traditional white collar jobs at risk. But as Rodrik (2018) points out, it is much more difficult to disentangle the effects of technical progress on real wages and employment from globalization effects causing the same movements.

There is now ample evidence that redistributive policies, including social protection expenditures appear to no longer harm growth prospects (Ostry et al. 2014) in recent years. Traditionally government expenditure, even for good causes, was felt to distort the economy away from efficient outcomes. Consequently, economic efficiency and equity considerations needed to be *separated*; moreover there was an efficiency–equity trade-off (see Okun 1975), empirical evidence for which seems to have weakened in recent years, linked among other phenomenon to new forms of market failure and extreme inequality.

This leads us to the political economy considerations, as the choice and implementation of policies by the state are fundamentally the outcome of strategic interaction between different factions with different interests. Downs' (1957) famous median voter theory suggests that the median voter's preferences prevail in a democracy. Meltzer and Richard (1981) suggest that increases in the size of government in democracies are due to median voter pressure for redistribution. This is either due to the extension of the franchise (which lowers the median voter's income relative to the national mean income), or when economic growth or transformation raises the average, or mean income above that of the median voter, implying greater inequality. The idea being that when the average voter becomes relatively (but not absolutely) poorer compared to the national average, he will vote for more redistributive policies.

Our present era, however, can be characterized by the steady loss of median voter power. The rise in inequality world-wide has meant that the average or mean income is rising faster than median income.[8] There has been a considerable shift in political clout away from the median voter to the policies that suit the super-rich, who compel the formulation of national policies to further the interests of the owners of internationally mobile capital and work skills. This manifests itself, chiefly, in the nation state feeling compelled to follow policies of fiscal austerity and wage compression, lest national participation in the globalized economy is jeopardized. This also frays the domestic social contract, and leads to the diminution of social protection. By contrast during the era of more limited globalization prior to the 1980s[9], the interests of the rich and the median income group did not necessarily conflict; it was an era of growing social protection, the provision of public services like education and health, employment rights, declining inequality and consensual democracy. The recent phenomenon of hyper-globalization (Rodrik 2017), defined to occur when the costs of further globalization in terms of the increased inequality outweigh economic benefits, produces a democratic deficit and a nasty backlash; this is what we turn to in the next section.

Illiberalism

Since the expression 'illiberal democracy' was coined by Fareed Zakaria (1997), a considerable number of scholars and commentators have drawn our attention to the recent rising tide of populist illiberalism and increased authoritarianism; Rodrik (2017), for example. The trend towards mainly right-wing populism is a feature mainly of developed countries that are established democracies; the authoritarian feature is mainly ascribable to developing countries. As Rodrik (2017) puts it, a liberal society is one where there is respect for minority rights, constraints on the executive (and legislature), with an independent judiciary, and there is respect for the rule of law. In a democracy there is an electoral process in place. A purely electoral process may, however, elect populist dictators or parties with scant respect for liberalism. The tyranny of the majority may engender illiberal actions, as described by classical liberals (Mill 1859; de Tocqueville, 1835).[10] Additionally, populist leaders (referred to as demagogues in Aristotle's Politics) can cause the degeneration of the polity.

Related to liberalism, are notions of the *liberal peace* that has important implications for both the internal workings of nations, as well as the relationships between countries. These ideas can be traced back to the work of Immanuel Kant, who in his essay on the Perpetual Peace (1795) argues that although war may be the natural state of man[11], peace can be established through deliberate design. This requires the adoption of a republican constitution simultaneously by all nations, which *inter alia* checks the war-like tendencies of monarchs and the citizenry; the resultant *cosmopolitanism* that emerges amongst the comity of nations would preclude war.[12]

Kant's (1795) thoughts provide us with information about the nature of the republican constitution that is of relevance to present day developing countries. First, observe the usage of the expression 'perpetual,' implying permanence as opposed to a transient or opportunistic peace. Secondly, and most crucially, Kant refers to the separation of powers between the executive and legislature; we may also add the independence of the judiciary. Thirdly, the stability of the peace depends upon the source of sovereignty or legitimate power within the nation. Although like all classical liberals Kant was not in favour of majoritarian democracy based on universal suffrage, he nevertheless points out that good governance provided by a dictator or an absolute monarch is inherently unstable as he or his successors face temptations to deviate from good government, and the assurance of good governance is more forthcoming in a system of power that is representative of the people. Central to the Kantian republican constitution is a system of checks and balances or a separation of powers which is precisely what populist politicians across the globe are so assiduously undermining.

Mirroring Kant's thoughts, is the contemporary philosopher, John Rawl's (1999) notion of peace between liberal societies, which he refers to as *peoples* and not states. Rawls' Law of Peoples is inspired by Kant's *foedus pacificum* and is termed a 'realistic utopia.' An ideal state is *reasonable*, even if in an imperfect world it is *rational* to deviate from such optima. He speaks of well-ordered peoples. These are mainly constitutional liberal democracies, which arrive at such a polity based on an idea of *public reason*. Public reason encompasses the realm of the political, and is not necessarily part of any comprehensive doctrine that individuals may believe in (for example religions or secular beliefs such as Marxism. In a well ordered society based on public reason human rights are respected, and the distribution of primary goods (a decent living standard, dignity, respect, and the ability to participate) for each citizen's functioning are acceptably arranged. Above all, the principle of *reciprocity* characterizes the determination and functioning of public reason, and its workings as a constitutional liberal democracy. This implies both tolerance of difference, and respect for all other citizens. Liberal societies only fight in self-defence, and invade to prevent gross human rights abuses and genocides. Within societies, the emphasis has to be on *tolerance*, akin to notions of liberty elucidated upon by Mill (1859), once again something that is undermined by populism.

The liberal peace has also an important economic dimension, a wheel greased by trade between nations promoting peaceful interaction, traceable to the work of Montesquieu (1748) and Paine (1791, 1792). To summarize the liberal peace, Gleditsch (2008) postulates a liberal 'tripod' where shared democratic values, economic inter-dependence and the common membership of international organizations together buttress the 'liberal' peace. The term 'liberal' inter-nationalist should be underscored in this regard, to distinguish this school of thought from 'realist' strands in political science and international relations, or for that matter neo-conservative thinking, which also advocate the joint merits of market capitalism and democracy.

Returning to the subject of the growing tide of populist illiberalism in our contemporary world, Rodrik (2018) suggests that the rise in populism coin-cides with hyper-globalization. The vote share of populist parties since 2000 in selected European and Latin American nations has exceeded 10% (Rodrik 2018, figure 1). Even in countries where the absence of proportional representation sidelines populist parties in legislatures, populist politicians can function within mainstream political parties. In Latin America, Venezuela for example, populism tends to be left-wing, harking back to the tradition of Peronism in the Argentina during the 1930s.

In Europe, by contrast, with the exception of some movements in Greece, Spain, and Italy, most populist parties are right wing. Although the support group for populism includes those impoverished by globalization, or millennials (youth) whose economic prospects are bleaker than for their parent's genera-tion, the phenomenon of mass migration in Europe has empowered right-wing demagogues who conflate the disadvantaging effects of hyper-globalization with immigration, especially Muslim immigrants (Rodrik 2018). Murshed (2011) outlines a theoretical model where a fear message is sent out by a populist right-wing politician. The potential vote bank must interpret the signal for what it is. Those with a negative experience of migrants (Muslims, Africans), those who are older, and especially those with less education often cannot separate the noise from the signal, and subscribe to the negative message. For the UK, for exam-ple, Becker, Fetzer and Novy (2017) indicate that the Brexit vote was greater in electoral districts with greater economic disadvantage after controlling for age and education. The right-wing politician simply uses the anti-immigrant, or Islamophobic stance as a ploy to get elected, but could have little intention of fundamentally rolling back the inequality that hyper-globalization has pro-duced. Indeed, it can be argued that in recent elections in the USA in 2016 (electing Donald Trump) and in the December 2019 election in the UK, the median American and English voter resounding voted to become poorer. The same can be said of the Indian elections of 2019 which kept Narendra Modi in power. Despite the nationalistic rhetoric, the ultimate outcome of the policies of these elected administrations will further immiserize the already poor in the United States, England, and India. Two explanations can be offered to rationalize this irrationality in the sense of Downs (1957). The first is cognitive dissonance and/or fooling. The second is to do with the 'memetic' viral spread of ideas[13], leading to identity based nationalism triumphing over economic interests in the minds of the median voter, who is willing to sacrifice economic interests[14] for the sake of nationalist identity based actions such as America First, Brexit, or the proscription of Muslims in India.

Indeed, it can be argued that it is no small coincidence that liberalism and democracy, as outlined above, co-exist (Rodrik 2017). The majority can always tyrannize the minority in purely elective democracies, especially in the absence of liberal constraints on executive power. Mukand and Rodrik (2020) analyse a model in which society is fragmented along three lines: there is a small elite who are characterized by their wealth, then a majority group that is poorer, and a minority who are differentiable from the majority by their different ethnicity. They use the model to distinguish between the rise of right-wing populism and left-wing populism. The former can be characterized by those that exploit cultural differences with the minority, and the latter by those that champion the cause of the relatively poorer. It may be that right-wing populism is more appealing to voters in societies experiencing greater immigration by those with different ethnicities and religion, and where the degree of social protection did not keep pace with the increase in inequality and marginalization due to policies of austerity. These members of the public conflate both economic and social problems with the influx of migrants, choosing to scapegoat them for nearly all of society's difficulties. Left-wing movements arise when the majority decide to stand up to the economic elites.

There are cultural and economic explanations for such behaviour (Rodrik 2019). The cultural explanation emphasizes alienation, exemplified not just in the fear of the 'other,' but also in inter-generational and educational divides (Norris and Inglehart 2019). Younger, more educated, and economically secure generations embrace more liberal and cosmopolitan values in contrast to older more conservative generations (Eichengreen 2018). The economic explanations point to trends in declining real wages, precarious employment, labour saving technical progress, trade shocks, re-location to low cost countries, declining social protection all of which culminate in economic inequality and insecurity for below median households (Autor et al. 2017; Becker et al. 2017 for example). Economic shocks such as financial crises lower trust in existing leadership (Algan et al. 2017). A realignment of politics away from the traditional left-right divide to an elite versus non-elite struggle creates vacuums that populists can fill (Gennaioli and Tabellini 2019). A sanguinary blow directed at established elites by supporting populists at the risk of further impoverishment is considered preferable to the centre-left redistributive agenda by many alienated, conservative and insecure voters.

In developing countries there has also been a rise in inequality, particularly in nations experiencing economic structural transformation. Associated with this is a rise in authoritarian tendencies, even in allegedly authoritarian developmental states, who employ elections and plebiscites. The V-Dem project produces several indices of democracy; see V-Dem (2017). Their definition of liberal democracy combines electoral democracy with a rule of law that involves respect for civil liberties, and judicial, as well as legislative, constraints on executive power. In the period between 2011 and 2016, more countries experienced statistically significant declines in this index than did the number which made progress in this respect. Most of these countries were in the developing world. If one looks at only the liberal component of the liberal democracy concept (excluding the electoral process), then 10 countries have made an advance, whereas 13 countries have experienced a decline in the liberal component between 2006 and 2016. When looking at the participatory principle of democracy, it is noteworthy that there has been a rise in the use of plebiscites and referenda in recent years (Altman 2017). This may, on the surface, appear as good news for democracy, but in reality this form of democracy is a tool utilized by authoritarian executives to override legislative and constitutional constraints on their exercise of power.[15]

Conclusions

I have argued that we need to broaden the focus of development studies towards addressing the wider problems associated with the twin scourges of inequality and illiberalism. Thankfully, this is occurring. In essence, the phenomena of inequality and illiberalism are inextricably intertwined; the former is a causal factor behind the emergence of the latter. A liberal and tolerant society in the sense of Kant, Mill or Rawls cannot exist under circumstances of extreme and growing inequality. The rise in inequality, accompanied by the falling standard of living for many amongst generalized increases in prosperity, coupled with the rise in insecure employment and falling social protection helped give rise to populist politics, as is now widely accepted. In connection with populism, I have attempted to reemphasize the point made by many scholars that liberalism and electoral democracy based on universal adult franchise need not always co-exist. Populism can, of course, take a variety of forms, but the common thread in all right-wing and authoritarian populism is a strident nationalism. I have included the growth of 'developmental' authoritarianism amongst the rulers of developing countries within the illiberal trend that we are witnessing.

The global trend in inequality that we have been experiencing for the last four decades, and especially since the advent of the new millennium, creates entrenched inequalities of opportunity that are unjustifiable on both the grounds of equity, as well as economic efficiency. In many nations of the global North, this trend also produces inter-generational inequality, as the younger generation are seriously disadvantaged compared to the older generations who had more opportunities to accumulate human capital and wealth during times when public goods provision was more plentiful, and real wages were on an upward trend. The principal factors underlying this secular trend towards increased inequality are long waves in real wage compression even for median income-earners and the creeping insecurity and informalisation of employment. This has been a long-term trend, and not just a feature of downturns in the business cycle. Another factor that has promoted inequality is the decline in social protection in developed countries during long phases of austerity, including the public provision of education and health services, factors that both enhance human capital accumulation, as well as reducing inequality. As Piketty (2014) emphasizes, inequalities in wealth are far more significant than inequalities in income because of the enduring nature of the former, implying that the long-term solution to inequality lies in wealth taxes.

Is economic globalization the chief culprit in causing the rise in economic inequality, and the consequent rise in populist-nationalist illiberal politics? It is true that increased international trade alters the functional distribution of income, and in recent decades this has meant the immiserization of unskilled labour. Financial globalization genuinely promotes inequality. The threat of relocation of production in an era of globalization also serves to emasculate the organized power of labour. More open economies are also more susceptible to external economic shocks, requiring them to have bigger governments providing greater social protection to cushion the citizenry against these shocks (Rodrik 1998). But another development, which helps to promote inequality and real wage compression: labour saving technical progress in manufacturing and also services. This tendency cannot be solely ascribed to globalization, and may take place independent of the extent of globalization. Criticism of 'hyper-globalization' emanates from both liberal and populist circles. Rodrik (2017) has pointed out the globalization trilemma, whereby the simultaneous achievement of national sovereignty, democracy, and hyper-globalization is impossible. He advocates a more diminished form of globalization such as that which existed during the Bretton Woods era, particularly during the 1960s and 1970s, when there were curbs on financial flows, and the rules governing the global economy were subject to the nation state's redistributive social contract. Populists too are wary of globalization; there is some evidence of de-globalization in the world economy (van Bergeijk 2019), highlighted by the aggressive trade policies of the USA since the inauguration of Donald Trump as President, and the COVID-19 pandemic is bound to disrupt global supply chains, forcing enterprises to rely more on domestic intermediate inputs.

At the time of writing the world is beset by the COVID-19 pandemic. Besides potentially causing the greatest recession since the Great Depression of the 1930s it has also exposed the extent of state failure in coping with this phenomenon. This painfully demonstrates the futility of the fiscal austerity pursued in the last two decades, which not only fanned inequality, but is also responsible for excessive COVID-19 fatalities. It is the disadvantaged all over the world whose lives and livelihoods are at the greatest risk. The question that remains is will the COVID-19 pandemic herald the demise of neo-liberal economic policies, just as the Black Death ushered in the demise of feudalism in 14th century England. History teaches us that this will not occur without a protracted struggle.

Notes

1 The doyen of this approach could be regarded as Amartya Sen; see for example Sen (1999)
2 Economic growth can both reduce or increase inequality, which usually refers to the distribution of income. Given a fixed standard of measuring poverty, economic growth will reduce poverty unless the distribution of income worsens, considerably.
3 In developing countries (mainly in Asia) that have experienced an increase in their export of unskilled labour intensive goods (such as garments) one would expect a rise in the remuneration of the unskilled relative to the skilled. But this has not happened because of the shortage of skilled personnel (less public education expenditure) and the huge numbers of unskilled workers coming from the hinterland; see Mamoon and Murshed (2008).
4 According to Piketty (2014), the alarming trend in inequality mainly stems from wealth inequality, as well as the income share of top groups relative to poorest groups. There is a tendency for the wealth to national income ratios to increase since the 1970s, having declined for a period prior to that; wealth, whose ownership is more concentrated than income, multiplies faster than wage income creating an ever widening gap between capital and labour, the biggest source of inequality.
5 A GINI coefficient is the most common metric utilised to measure inequlity. It is the the sum of differences of all individuals or groups from the average or mean. It ranges from 0 to 100, with the former implying perfect equality and the latter indicating that one person or group has all of society's endowment. Thus, increases in the GINI coefficient indicate rising inequality.
6 This is because the extremely wealthy demand a high return to their financial investments, and the financial debt burden of the relatively poor, if securitized, can make economies more prone to financial crises, which in turn can cause major recessions.
7 The recent COVID-19 pandemic may slow down this phenomenon.
8 Rodrik (2017) cites the work of Mizruchi (2017) who argued that in the early postwar era (1945 to the 1970s) the corporate elite in the United States exhibited ethical considerations of civic responsibility and enlightened self-interest.
9 Authors, such as Rodrik (2017) refer to this period as the Bretton Woods era (1945–1973).
10 Arguably, the early constitution design of the United States with its checks and balances, separation of powers, as well as a bill of rights was designed to prevent such outcomes.
11 Akin to Thomas Hobbes' conception of the non-contractual 'state of nature.'
12 This could also create a confederation of nations with common values, such as with the European Union's *acquis communautaire*.
13 This is like the viral spread of ideas, attributable to Richard Dawkins; see, Dawkins (2006), for example, where he utilizes this term to describe the spread of religion. A meme is an idea that can go viral, and whose spread may be facilitated by widespread internet access.
14 Here it can be argued that this median voter, especially in the USA or the UK, is middle-aged or older, and may be someone who is acquired a comfortable degree of wealth, and whose nationalistic voting preferences harm the economic prospects of the younger generation, his children or grandchildren.

15 The British Prime Minister between 1945–1951, Clement Attlee spoke of referenda in the following terms: "I could not consent to the introduction into our national life of a device so alien to all our traditions as the referendum which has only too often been the instrument of Nazism and fascism," https://www.thelondoneconomic.com/politics/this-clement-attlee-quote-on-referendums-is-going-viral/02/09/, accessed 20th April 2020.

References

Algan, Y., Guriev, S., Papaioannou, E. and Passari, E. (2017). The European trust crisis and the rise of populism. *Brookings Papers on Economic Activity Fall*, pp. 309–382.

Altman, D. (2017). The potential of direct democracy: A global measure (1990–2014). *Social Indicators Research*, 3, pp. 1–21.

Aristotle. (1905). *Politics*. Oxford: Clarendon Press.

Autor, D., Dorn, D., Hanson, G. and Majlesi, Kaveh (2017). Importing Political Polarization? The Electoral Consequences of Rising Trade Exposure. MIT Working Paper.

Becker, S., Fetzer, T. and Novy, Dennis (2017). Who voted for brexit? A comprehensive District-level analysis. *Economic Policy*, 32(92), pp. 601–650.

Dabla-Norris, E., Kochkar, K., Suphaphiphat, N., Ricka, F. and Tsounta, E. (2015). *Causes and Consequences of Income Inequality: A Global Perspective*. IMF Staff Discussion Note.

Dawkins, R. (2006). *The God Delusion*. Bantam Books.

De Tocqueville, A. (1835, 2000). *Democracy in America*. Chicago: University of Chicago Press.

Downs, A. (1957). *An Economic Theory of Democracy*. New York: Harper and Row.

Eichengreen, B. (2018). *The Populist Temptation: Economic Grievance and Political Reaction in the Modern Era*. Oxford: Oxford University Press.

Furceri, D., Loungani, P and Ostry, J. (2017). *The Aggregate and Distributional Effects of Financial Globalization*. Washington: International Monetary Fund.

Gennaioli, N. and Tabellini, G. (2019). *Identity, Beliefs, and Political Conflict*. CEPR Discussion Paper No. DP13390.

Gleditsch, N.P. (2008). The liberal moment fifteen years on. *International Studies Quarterly*, 15(4), pp. 691–712.

Goda, T. (2014). Global trends in Absolute and Relative Wealth Concentrations. CIEF Working Paper, 14-01-2014, ile:///C:/Users/ab2380/Downloads/SSRN-id2395003.pdf, accessed 6th March 2018.

Huntington, S. P. (1993). *The Third Wave: Democratization in the Late Twentieth Century*. Norman, OK: University of Oklahoma Press.

Jayadev, A., Lahoti, R. and Reddy, S. (2015). Who Got What, Then and Now? A Fifty Year Overview from the Global Consumption and Income Project. Working Paper 10/2015, The New School for Social Research.

Kakwani, N. and Pernia, E. (2000). 'What is pro-poor Growth.' *Asian Development Review*, 16(1), pp. 1–16.

Kant, I. (1795). *Perpetual Peace and Other Essays on Politics, History and Morals*. Reprinted 1983. Hackett Publishing.

Lakner, C. and Milanovic, B. (2015). Global income distribution: From the fall of the Berlin Wall to the Great recession. *World Bank Economic Review*, 30(2), pp. 203–232.

Mamoon, D. and Murshed, S.M. (2008). Unequal skill premiums and trade liberalization: Is education the missing link? *Economics Letters*, 100(2), pp. 262–266.

Meltzer, A. and Richard, S. (1981). A rational theory of government. *Journal of Political Economy*, 89(5), pp. 914–927.

Mill, J.S. (1859). *On Liberty*, Longman, Roberts and Green.

Mizruchi, M. (2013). *The Fracturing of the American Corporate Elite*, Harvard University Press.

Montesquieu, C-L. de. (1748). *De l'Espirit Des Lois*. Reprinted 1979. Flammarion.

Mukand, S. and Rodrik, D. (2020) The political economy of liberal democracy. *Economic Journal*, https://doi.org/10.1093/ej/ueaa004.

Murshed, S. M. (2011). The clash of civilizations and the interaction between fear and hatred. *International Area Studies Review*, 14(1), pp. 31–48.

Norris, P. and Inglehart, R. (2019). *Cultural Backlash: Trump, Brexit and Authoritarian Populism*. Cambridge: Cambridge University Press.

Okun, A. (1975). *Equality and Efficiency: the Big Trade-Off*. Washington: Brookings Press.

Ostry, J., Berg, A. and Tsangarides, C. (2014). Redistribution, Inequality and Growth, IMF Staff Discussion Note.

OXFAM (2016). Wealth: Having it All and Wanting More, https://www.oxfam.org/sites/www.oxfam.org/files/file_attachments/ib-wealth-having-all-wanting-more-190115-en.pdf, accessed 23rd January 2016.

Paine, T.. (1791, 1792). *The rights of Man*. In: M. Philip (ed.), Oxford: Classics.

Piketty, T. (2014). *Capital in the Twenty-First Century*. Cambridge, Mass.: Harvard University Press.

Rawls, J. (1999). *The Law of Peoples*. Cambridge, Mass.: Harvard University Press.

Rodrik, D. (1998). Why do more Open countries have bigger government. *Journal of Political Economy*, 106(5), pp. 997–1032.

Rodrik, D. (2017.) *Straight Talk on Trade: Ideas for a Sane World Economy*. Princeton University Press.

Rodrik, D. (2018). Populism and the economics of globalization. *Journal of International Business Policy*, https://doi.org/10.1057/s42214-018-0001-4.

Rodrik, D. (2019). "What's driving populism?," *Project Syndicate*, 9 July.

Roemer, J. E. (1998). *Equality of Opportunity*. Cambridge: Harvard University Press.

Sen, A.K. (1999). *Development as Freedom*. Alfred Knopf.

Standing, G. (2011). *The PrecaRiat: The New Dangerous Class*. Bloomsbury Academic.

Stiglitz, J.E. (2012). *The Price of Inequality: How Today's Divided Society Endangers Our Future*. New York: Norton.

Sumner, A. (2016). *Global Poverty: Deprivation, Distribution, and Development Since the Cold War*. Oxford: Oxford University Press.

Van Bergeijk, P. A. G. (2019). *Deglobalization 2.0: Trade and Openness During the Great Recession and the Great Recession*. Edward Elgar.

V-Dem Institute, Varieties of Democracy. (2017). Annual Report, https://www.v-dem.net/en/news-publications/annual-report/, accessed 9th March 2018.

Zakaria, F. (1997). The rise of illiberal democracy. *Foreign Affairs*, 76(6), pp. 22–43.

5

GENDER, CLIMATE, AND CONFLICT IN FORCED MIGRATION

Tricia Glazebrook

Introduction

Living in Syracuse, NY, I was apprenticed to a master drummer from Ghana. Some hundred or so 'lost boys' of the Sudanese genocide in Darfur were re-located to Syracuse, and we occasionally played for them. Some were happy in their new situation and open to conversation. Most of them had walked hundreds of miles across a desert before reaching a camp. Some did not talk and their face never changed. Sometimes they responded to the drums, but would not interact with anyone. They must have survived unspeakable situations as child soldiers, kidnapped, addicted to heroin, and trained as killers. Twice, with those I had talked with several times, I eventually asked, where are your sisters? They said, 'they stayed in the village because it was safer for them.' They would no longer meet my eyes. They may not have known what happened to their sisters, mothers, and aunts but they knew that if women and girls survived the initial attack, they were either raped and killed on the spot, or taken to do domestic and often sexual work for the Janjaweed. The Janjaweed were originally a militia of rebel groups until the Sudanese government saw the value in paying them to commit genocidal murder while keeping the government at a distance (Glazebrook and Story 2012).

This chapter examines the role of neoliberal, market-based, global economics in driving people to flee their homeland and in managing their subsequent experience, with especial attention to what women and girls face when in forced migration. After provision of historical and social context concerning refugee policy, gender and sexual violence, and oil and conflict, popular perception of neoliberal economics is critiqued as false. Finally, a market-based solution to current refugee crises is presented and rejected. Bio-logics of care are described as an alternative that precludes the dehumanization of women and girls in forced migration that reduces them to bodies available for sexual exploitation in a process called 'abjectification, i.e. the rendering disposable of a living being. First, however, what it means in international law to be a 'refugee' and why gender is a crucial axis of analysis are explained.

According to the United Nations High Commissioner for Refugees (UNHCR), 70.8 million people are forcibly displaced globally (UNHCR 2001-20). Of these, 3.5 million seek political asylum, 43.1 million are displaced in their homeland, and 25.9 million are refugees. The UNHCR defines 'refugees' as anyone who, 'owing to well-founded fear of being persecuted for reasons of race, religion, nationality or political opinion, is outside the country of his nationality [or former habitual residence] and is unable or ... unwilling to avail himself of the protection of that country' (UNHCR 2010). The 1951 Convention on the Status of Refugees was developed to deal with people fleeing events in Europe in connection with the Second World War. A subsequent 1967 Protocol expanded the Convention beyond 'any geographic limitation' (UNHCR 2010), but it still considers refugees to be forced to flee from persecution, human rights (HR) violation or conflict.

Gender is crucial to the analysis. Betts and Collier (2017) have little to say about women or climate-forced migration, despite the fact that for 1.4 billion women on the planet, mostly in the global South, agriculture is their primary livelihood (FAO 2012). Moreover, 70% of the 1.3 billion people living on less than $1 a day are women, and women are 60% of the world's chronically hungry (Roberts 2009). Climate change exacerbates these situations and makes women especially vulnerable to climate-forced migration. The purpose of this chapter in engaging gender is not to exploit women's vulnerabilities and traumas through theoretical argument, but to raise awareness of the harms women and girl refugees are subjected to as a call for their protection and support, and to show that such gender-abjectification is just as enabled by contemporary global economics as is exploitation of resources that are similarly instrumentalized. 'Androcene' displaces 'Anthropocene' in order to recognize the phallic logic underwriting contemporary global economics in the high-carbon epoch marked permanently in the geological record (Glazebrook 2016). This chapter is dedicated to the many women in forced migration whose life has been damaged, stolen, or lost through androcentric exploitation.

Historical and social context

Three factors need explanation to preface the policy-gender-capital interconnection under scrutiny in this chapter. This section lays them out in brief histories of international refugee policy, women's and girls' vulnerability to sexual exploitation in flight and in camps, and the function of oil and conflict in global economics.

Refugee policy

Beyond conflict, natural disasters, and climate conditions that render locations unlivable force people to migrate. Climate is the oldest driver of migration, if the 'out of Africa' hypothesis of global human spread is correct (Jensen 2007). The definition of 'refugee' in international humanitarian law above, however, leaves people forced into migration by ecosystem breakdown in a legal void, neither 'refugees' under the 1951 Convention, nor 'migrants' relocating by choice. They accordingly have no protections afforded refugees or migrants.

The United Nations Framework on Climate Change recognized this problem in the 2010 Cancún Adaptation Framework that called for '[enhanced] understanding, coordination and cooperation with regard to climate change induced displacement, migration, and planned relocation' (UNFCCC 2010). The UNHCR took up this cause and by late 2012 the Nansen Initiative established a set of principles to provide a global framework for treatment of people in flight from climate conditions (Nansen Initiative 2015). The Initiative remained controversial but was able to identify blockages in the process, including reluctance to acknowledge the reality of anthropogenic climate change (Kälin 2012) and concern that the UNHCR was inflating its interests by moving from humanitarian aid to a legal 'protection' mandate (McAdam 2016). The UNHCR reduced reactions to 'climate change' by repudiating the phrase 'climate refugee' on grounds that no such thing exists in international law, and simply kept going because the need is exactly to replace 'refugee' with 'people in forced migration,' and humanitarian aid has always entailed working toward international humanitarian law.

In truth, climate, disaster, and conflict virtually never happen in isolation. Disasters can be climate induced, e.g. fires and extreme weather events, or have climate impacts, e.g. cooling effects of volcanic ash. Likewise, climate stress can cause conflict when resources become sparse, e.g. the 2003–2005 Darfur genocide began when the Sahara creeping south pushed Arab herders onto non-Arab farmland (Johnson 2011; Glazebrook and Story 2012). Similarly, the Rwandan Banyamulenge, who found themselves in eastern République Démocratique du Congo (RDC) after the partition of Africa, are currently under assault as 'foreigners' in disputes over grazing rights on land made less productive by climate change (Glazebrook 2020). Much conflict, regardless of explanatory narrative, is resource acquisition, especially land-grabbing for agriculture and pastoralism, but also, as discussed below, oil.

Connections between climate, disaster, and conflict have led to understanding of 'nexus dynamics' in which flight is more complex than recognized in traditional definitions. After the Nansen Initiative came the 2016 Platform on Disaster Displacement (PDD 2016), that provides a Protection Agenda for preventing and preparing for displacement, and responding to forced migration, followed by the 2018 Global Compact on Refugees (GCR 2018). Phrasing in terms of disaster avoids tensions around 'climate' discourse, and many climate harms are disasters, though perceptions of 'disaster' vary. Droughts that collapse food security are slow-onset, while 'disasters' can be perceived as sudden, catastrophic events. Defining 'refugees' not in terms of the driver of their flight but as persons in forced, rather than voluntary, migration resolves this issue and is consistent with the original 1951 articulation of persons fleeing for survival.

Gender and sexual violence

In 2008, the UN Security Council noted that women and girls are targeted for sexual violence as a tactic of war (UNSC 2008). As noted by Major General Cammaert, a retired Dutch officer in peacekeeping, 'it is now more dangerous to be a woman than to be a soldier in modern conflict' (Roberts 2009). In 2008, moreover, women's poverty and lack of power were identified as factors that make women vulnerable to climate migration (Roberts 2009). These two insights were not connected, however, in refugee policy. Ten years later, disasters and climate both were accepted as factors in the nexus dynamics of forced migration. The 2018 program of action for GCR implementation explicitly and repeatedly responded to women's and girls' vulnerability to sexual violence by promoting gender equality and calling for the end of human trafficking and all forms of sexual and gender-based violence, exploitation, and abuse. It gave this thematic account from the perspective of women's empowerment through protections, education, and 'decent work' (GCR 2018). This uniquely gender-sensitive policy document not only identifies and responds to the threat of sexual violence toward women and girls but recognizes them as autonomous agents capable of full participation in society.

Unfortunately, however, Weerasinghe's 2018 report to the UNHCR addressing the 'nexus dynamics' knowledge gap in understanding migration reverts back to failure to recognize women's unique challenges in nexus dynamics. Weerasinghe says almost nothing about gender, using the word 'women' only eight times, 'girls' only once in a footnote, and 'female' three times in 'female-headed household' (Weerasinghe 2018). This is especially disturbing given that the report's intention is to provide recommendations 'to strengthen implementation of refugee law-based international protection' (Weerasinghe 2018). For a consultant providing recommendations to protect people in flight, absence of dedicated, thematic discussion of gender is an egregious shortfall returning to the gender-blindness that risks women's and girl's safety, well-being, participation, autonomy, and development in flight. After all, in 2009, for example, of the 50 million people displaced internationally, approximately 80% were women and children (Roberts 2009). Not addressing protection of women and girls from sexual abuse exacerbates their situation in many ways, including for example, poor design that precludes women's access to camp toilets at night because of the risk of assault while walking to their location.

The sexual exploitation of women and girls in forced migration violates at least nine Human Rights (UDHR 1948, Articles 3–7, 12, 16, 22) yet receives disproportionately little policy attention, and the durability of explicit discussion is fragile. Forced migration of women and girls provides simpler than normal (and more) opportunity to sexual abusers with less accountability, in a larger global context where gender-based sexual violence is normalized in widespread contemporary 'rape culture' (Griffin 1971; MacKinnon 1999). In the Androcene, women and girls are dehumanized in their reduction to an exploitable resource that is disposable. She is abject—etymologically, 'thrown away'—once she serves her function. For example, in Jordan's Zataari refugee camp, young women are sold by fathers or marriage-brokers, often women, into 'pleasure marriages' with men from nearby gulf states who often simply abandon the young woman somewhere in Jordan after a few days so she must find her way back to the camp alone (Long 2013). Regardless of consent conditions, that are at best blurry, her autonomy is lost in her dehumanization into a body that serves only instrumental value. This principle of reduction to exploitable resource is fundamental to global economic systems.

Oil and conflict

Global greed currently manifests as corporate greed in the uniquely lucrative oil and gas industry that mobilizes governments at best as its tool, and at worst as its weapon in the face of resistance by indigenous and other cultural groups protecting their land and water resources because agriculture is their primary livelihood. Women are largely invisible in the global economy insofar as much of their work is unpaid domestic labour, including crop production to provide family food security, that has no place in traditional economic reckonings (Waring 1988). Yet the global South cannot feed its people without women's subsistence crops. In Africa, a policy contradiction is well underway as countries develop fossil fuel resources to pull themselves out of poverty on one hand while on the other, developing policy to support women's substantial contribution to food security that is in decline because of climate impacts (Glazebrook 2011). Fossil fuels are accordingly central in the nexus dynamics of forced migration.

Fossil fuel interests are also at the heart of global conflict. For example, Dick Cheney was focal in planning and launching Operation Desert Storm, the 1991 invasion of Iraq after Iraq's invasion of Kuwait, and a number of controversies connect him with the later 2003–2011 U.S. Iraq War when he was U.S. Vice President (VP). He gave up his position as chairman and CEO of Halliburton Company, the world's second largest oil field service multinational corporation, in order to take up the VP position, and apparently received a $33.7 million USD separation package (Finnegan 2000) from Halliburton during the 2002 presidential campaign, along with almost $400,000 USD in deferred compensation and stock options. Just before the invasion of Iraq, while Cheney was VP, Halliburton received a $7 billion USD oil extraction contract with Iraq for which it was the only company allowed to bid (Corbin 2008). Months before the U.S. military bombed Iraq, the Department of Defence was secretly working with Halliburton on a deal to give them control over Iraq's oil fields, according to Halliburton executives (Leopold 2003). Halliburton presumably profited from Iraqi oil, and Cheney certainly profited well from his relationship with Halliburton. This example clearly exposes the synchronicity of individual and corporate greed with government promotion of corruption and violence for wealth acquisition in a system that was at best opportunistic and at worst intentional in causing deaths, displacement, and other horrors of war.

The larger picture is not, however, just corruption and individual profit. Rather, Cheney's and Halliburton's exploitation of Iraqi oil is symptomatic of deeper conflict generated by the role of oil in global economics, known as the Petrodollar because of long-standing U.S. influence on oil prices. This influence was undermined by oil in the 1960s, when the Organization of the Petroleum Exporting Countries (OPEC) was formed whose member-states together controlled 40% of global oil production (OPEC 2020). The U.S. has consistently considered OPEC an economic threat. For some time, however, the global economy was also underpinned by the Eurodollar market in which the wealthy from around the world stored their money in U.S. dollars outside the U.S. By 1985, 75% of U.S. dollars were in foreign accounts outside U.S. control, thereby challenging U.S. capacity to influence global economics through currency exchange (Vatter and Walker 1995). Fracking, however, rapidly strengthened U.S. influence on oil prices at the centre of global economics as, for example, by 2019, the U.S. was producing more than 12 million barrels per day (Sharma 2020), despite OPEC's control of almost 75% of reserves and 42% of production. In 2016, however, OPEC had created OPEC+ to amalgamate with other high oil-producing nations, including Russia, that together control over 50% of global oil supplies, and about 90% of proven reserves. As of December 2019, global leaders in oil production are the U.S., Russia, and Iraq, in that order, though Saudi Arabia, Russia, and Iraq lead in export volumes (Sharma 2020). This emergence of states into oil development that seem willing to collaborate with OPEC threatens once again to displace newly recovered U.S. influence on global economics.

In early 2020, the U.S. killed Iranian Major General Qasem Soleimani after several disagreements over U.S. incursion of intelligence-collecting drones into Iranian territory and Iranian downing of a U.S. drone in June 2019 (Glazebrook 2019). MG Soleimani's murder has been identified as a 'worrying picture' of 'a desperate U.S. lashing out' at a world that seems to favour collaborative oil management over single-state economic power: 'the Petrodollar ... [grants] the U.S. a monopolistic position from which it derives enormous benefits...To threaten this comfortable arrangement is to threaten Washington's global power' (Pieraccini 2020). Emergence of Russia, China, Iraq, and Venezuela, for example, into global oil economics plays an increasing role in global markets, especially in consolidation through an organization such as OPEC. Venezuela, Russia and Iran's oil and gas holdings can increase OPEC's influence. These countries have 'elevated relations' with China, especially Russia that wants to consolidate with China allegedly to grow the Eurasian supercontinent peacefully (Pieraccini

2020) The remaining countries holding the majority of oil and gas reserves are Iraq, Qatar, and Saudi Arabia that could be swayed to China-Russia solidarity for both military and energy reasons, i.e. U.S. aggressions in the region, including the 2003 Iraq invasion and current drone assaults on Yemen. In less than one month in 2017, For example, forty strikes allegedly based on shaky intelligence killed hundreds of non-combatants in Yemen, including children (Glazebrook 2019).

Concerning the 2003 invasion of Iraq, moreover, the U.S. Ambassador to Iraq, April Glaspie, told Saddam Hussein, allegedly because she was so instructed, that the U.S. had no opinion on Arab-Arab conflicts such as the disagreement with Kuwait, and the State Department also told him that Washington had 'no special defense or security commitments to Kuwait' (Walt 2011). It seems Iraq was set up to invade Kuwait so the U.S. could in turn invade Iraq and take control of its oil resources. There is therefore good reason to believe that in the case of both the 2003 Iraq invasion and the killing of Soleimani, a message was being sent that the U.S. is not prepared to relinquish its singular authority over oil-based global economics. As a corollary, U.S. benefit from the Petrodollar potentially explains its on-going climate denial.

As a small group of oil company executives and politicians reap massive profits off oil revenues, and nation-states battle for control of the global economy through oil, millions of people lose virtually everything. Antonio Guterres, then UNHCR, reported in February 2017 that 2 million people had fled Iraq (NPR 2007) in consequence of the 2003–2011 conflict and another 1.7 million were internally displaced (O'Donnell and Newland 2008). The death toll of civilians was over 100,000 (BBC News 2011). That is, in contemporary global economics, profit trumps life.

Nexus dynamics of refugee policy, sexual violence, and oil conflicts

In summary, this section first tracked the evolution of international refugee policy from limited understanding of refugees as people forced to flee conflict and persecution to broader perspectives of people forced to leave their homeland by the nexus dynamics of conflict, violence, disaster, and climate. The second part uncovered particular risks to women and girls of sexual violence and exploitation during flight and a blind spot in policy that obliviates their protection from such harms, that were both attributed to a global phenomenon of 'rape culture' that dehumanizes women and girls as exploitable resources. The third part identified oil-based conflicts as struggles in global economics that evidence the priority of profit over life. That is, contemporary extreme capital accepts the military doctrine of 'double-effect,' i.e. justification of a secondary impact that is not intended but knowingly cannot be avoided in a necessary action, because extreme capital understands the meaning of human existence to be individual accumulation of private wealth.

This section accordingly concludes that in a second kind of nexus dynamics—capital, oil, and global economics—humanitarian crises and atrocities are collateral damage in battles of wealth and power. Within this egregious incapacity for moral awareness, a third set of nexus dynamics—capital, gender bias, and policy—works internationally to force women and girls into migration that exposes them to dehumanizing objectification and abjectification in which they are reduced to the exploitable resource of their body with no regard for their well-being, autonomy, or survival. The shared factor across all three nexus dynamics discussed in this paper is capital. Given that overwhelmingly, oil executives, high-level politicians, and sexual abusers are men (also when men are the object), one can conclude not that all men are profiteers and abusers, but that a phallic logic of wealth and power drives profiteering and sexual abuse.

Popular perception and critique: The free market and neoliberal economics

Many popular conceptions are false. The 2003 Iraq invasion was based on false connection by Cheney of Hussein and Al-Qaeda in the 9/11 attack, though the 911 Commission and the intelligence community reported no such connection (Pincus and Milbank 2004; Waas 2005). The murder of MG Soleimani was portrayed as a response to aggressions. Both have been argued above to be strategic actions to support U.S. influence on global, oil-based economics. Against popular climate denial in the U.S., climate impacts on agriculture have been shown to be especially damaging to women's food security practices that put hunger at the core of the nexus dynamics of forced migration. Against portrayal of women as passive victims, active rape culture has been argued to expose women in forced migration to sexual violence through dehumanization to a body that is available for exploitation. These misconceptions are founded on false assumptions about global economics.

Popular neoliberal perception is that economics functions best as a 'free market' in which supply and demand interact to create an 'invisible hand' that balances economies by driving price up or down when a commodity is sparse or plentiful. Policy interventions, e.g. price controls, rationing, or rent control, disrupt this balance by creating excess demand that raises prices beyond the budget of the poor, or excess supply that lowers prices and drives companies into bankruptcy. Rationing and supply caps also prompt black market exchange. No market is, however, in actual fact 'free.' The obscure if not 'invisible' economic hand globally is the hand of governments and cartels through trade agreements, and land and oil appropriation, or sanctions such as embargos and tariffs.

Even what are commonly understood as well-intended interventions into a country's economics cause harm in neoliberal contexts of global finance that create 'maldevelopment' (Mies and Shiva 1993). Aid programs create mass debt to the World Bank and International Monetary Fund, and attach unwelcome conditions. Jamaica's entry into the global economy, for example, devastated its industry and agriculture (Black 2001); unemployment was still almost 10% in 2019 (Statista 2020) despite GDP growth from $3.5 billion USD in 1992 to 15.7 billion in 2018 (World Bank 2020). By favouring traditional metrics such as GDP, neoliberal approaches do not actually indicate lived experiences of poverty. For example, land appropriation for corporate agriculture looks like development because GDP goes up; but the woman previously farming that land for subsistence has lost her food source and livelihood. That is, economic success is perceived on the basis of capital outcomes rather than ecosystem and human well-being. Accordingly, governments are at best the tools of capital, at worst, its weapon.

Emerging alternatives

This section first assesses and finds wanting an emerging, popular market-based solution for the problem of what to do with people in forced migration, followed by suggestions for establishing conditions that would displace phallic logics of capital through care-driven bio-logics.

Special economic zones

One of the most significant problems for a person in forced migration is livelihood. Those who enter camps receive a modicum of support for housing and basic needs but are usually denied the right to work, even after years or decades of residency, and so are caught in a kind of limbo. Those who choose to take their chances in a city face no support but can seek employment, usually in the informal economy that exposes them to abuses such as denial of pay, beating or sexual violence, against which they have little recourse, and poor labour conditions that may be unsafe or unhealthy. Both these options typically make prostitution a woman's best option. That is, capital enables a large number of women's participation only through sexual disciplines of femininity she would otherwise reject.

Betts and Collier (2017) propose Special Economic Zones (SEZs) as a market-based solution for everyone. SEZs are production areas where outputs are not subject to tax that are intended to create employment and trade capacity, and promote investment. An SEZ in a camp allows employed residents to regain autonomy, self-esteem, and hope for the future through earning an income. The host country gains revenues that can help support camp running costs. The origin state receives returning citizens post-conflict that have developed skills to participate effectively in rebuilding society. The global North benefits from reduced aid contributions, untaxed access to affordable goods, and an alternative to opening borders. SEZs promise an excellent system for turning hardship into revenue. Women in particular benefit from an alternative to prostitution.

Yet SEZs collect together automated, factory-style, industrial production in which line workers need only minimal training for quota-based repetitive work.

The idea of beneficial training assumes camp dwellers have no developed skill set, though many would not benefit from training that might draw them away from more specialized work requiring significantly more training and education, e.g. medicine or law. Working conditions are also a concern. SEZs make a country more globally competitive by producing and trading goods at lower prices, and can allow industry to exploit tax breaks without necessarily providing the promised employment or export earnings. They are unsustainable if labour costs rise, and they are considered by many economists only to be successfully competitive in specific conditions over limited time (Hamada 1974; Madani 1999; World Bank 1992, 2008). In the face of shortfalls, the easiest approach to remaining competitive is cutting costs, which often means paying employees less and keeping overhead at a minimum. That is, workers pay the cost of SEZs staying competitive by labouring in poor conditions for minimal pay. Workers can in fact find themselves exploited in 'sweatshop' conditions with little autonomy and poor access to resources, while the only real change from their camp situation is transition into forced labour bordering on slavery. Such conditions enable women's and girls' sexual exploitation by management and other workers. Neoliberal approaches are accordingly likely to fail because they are structured by and aimed at capital logics of profit rather than humanitarian logics of care.

Bio-logics and logics of capital

Bio-logics are life-oriented rather than profit-oriented. Bio-logics are a way of thinking but also an ontology of place and practice that has no need to get rid of capital as a resource that can function more equitably in systems that prioritize its social function. That is, capital is a reasonably effective system for organizing societies, both socially and structurally. But it is not neutral and, when reduced to private wealth accumulation over societal relationships in functional communities, it becomes a disastrous driver of politics corrupted to wealth and power. In contemporary economics, people, ecosystems, and co-inhabitant species are reduced to their instrumental value as tools and resources to be exploited. In bio-logics, capital appears as the tool rather than its own apotheositic end-in-itself, and people, species, and ecosystems inter-relationally enable life. Current logics of capital need shifting from within away from profit-oriented private interests and into corporate accountings of social costs and benefits. That is, externalities are more important in community and ecosystem thriving, and therefore in corporate accounting, than profits because profits are the zero-sum gain in resource and labour exploitation. Bio-logics accordingly both make, and are guided by the capital-supported goal of opening spaces for thriving, rather than the politico-economic capital goal of individual accumulation of private wealth. Biologics solve an inverted trolley problem in which a more equitable system diverts profits from the few to the many, and practically, in order to support long-term system thriving, draws from ecosystems systems sustainably which in turn reduces demands on labour in corporations owned by workers rather than share-holders.

For example, in the face of scientists throughout the globe overwhelmingly agreeing that human release of greenhouse gases into the atmosphere by burning fossil fuels is changing the climate in ways that are already in many places catastrophic and will become in the not too distant future irreversible, bio-logics favour transitioning to renewables. Bio-logics recommend shifting massive subsidies currently paid to oil companies to investment in carbon-neutral energy sources, and disablement of systems that enable corporate donations to trump public interest and voter voices. Bio-logics dissolve the corporate charter, hold executives accountable for corporation-caused ecosystem and human damage by referring them to the International Criminal Court (ICC) in the Hague as agreed in the Rome Statute, and confiscate stockholder profits from such disasters to fund the ICC and provide loss-and-damage support to affected populations.

Bio-logics also support women's reproductive autonomy because pronatalism contradicts climate adaptation, but also as recognition justice of women's right to bodily autonomy, and ability to make decisions based in care, as they often do daily for the lives around them, including, as farmers and pastoralists, the life of their animals. It is not that women inherently 'care' but that women's labour in inherently care-work, aimed at reproducing the material conditions of daily living, and virtually always unpaid, even if a woman is also otherwise employed. When women pass care-labours of childcare and housekeeping to another, it is most often to another woman. Women's work is liminal and straddles revenue generation and care in self-relevant and changing proportions. Bio-logics are not based on presumption that all women are mothers but the universal reality that every person's origin and first environment is a mother. The ontology of bio-logical gender awareness of its very nature precludes rape culture. That is, bio-logics are yonic care-cultures in contrast with phallic logics of exploitation, domination, and profit.

Kalle Lasn once made a short television advertisement exposing the logging industry in British Columbia, Canada as unsustainable (Lasn 1999). No broadcasting corporation would air it. Eventually an executive explained that they did not support special interests. The executive was blind to the fact that every advertisement they run is a special interest. Bio-logics see that phallic logics of exploitation are inherently destructive, and can be shifted to care-logics through public awareness-raising, peaceful protest, and education currently practiced by children and youth such as Greta Thunberg on climate, and Emma Gonzales and David Hogg on gun control who very rapidly changed many minds to new logics.

Conclusions

This chapter first argued in three parts that humanitarian crises leading to forced migration are collateral damage in battles of wealth and power within which the combination of capital, gender bias, and policy works internationally to force women and girls into migration and dehumanizes them into resources available to men for sexual exploitation. Given male domination of the global economy, conclusion was drawn that a phallic logic of wealth and power drives profiteering and the rape culture that normalizes the sexual abuse of women.

The free market was then argued to be a defunct and duplicitous illusion that enables neoliberal, market-based approaches in global management that assume the pursuit of wealth to be a human right and entitlement. Yet only the privileged can pursue wealth freely, and do so at the expense of others. The fundamental challenge of the 21st century has already shown itself as the need to address global inequities between North and South; for example, of the world's 795 million people going hungry in 2015, 780 million—over 98%—were in the global South (FAO et al. 2015). This massive inequity in global hunger has for more than a decade been exacerbated by global carbon emissions enabled by the oil industry that largely controls the global economy and is at the heart of much corruption and conflict.

The final section first examined the market-based solution to the current global refugee crisis of creation of Special Economic Zones to employ residents housed in camps to the benefit of the person, the host and origin country, and the global North. Argument was made that such ventures exploit labour and degrade working conditions in order to maximize profits. These factors, as well as exposure for women and girls to sexual abuse, do not improve the camp-residents' situation but easily make it worse. Finally, argument was made for bio-logics over capital logics of profit, power, and exploitation. Bio-logics were shown to be yonic and life-oriented care logics, and bio-logic practices were described as advisory suggestions for initiating change.

Together, these sections show that sexual violence against women in forced migration lies at the heart of: the nexus dynamics of gender blindness and fragile gender awareness in policy, despite the best efforts of the UNHCR; global economic wealth and power struggles, largely fought over oil that causes climate change and forces migration; and, rape culture that normalizes women's and girls' reduction to an exploitable sexual resource. What is needed to protect women in forced migration from sexual violence is change from phallic logics of exploitation and profit to bio-logics of care and life. I hope to have shown that transition from fossil fuels and from women's abjectification are deeply related transitions, and that what is needed is not just transition, but *a just transition*.

References

BBC News. (2011, December 14). Iraq war in figures. https://www.bbc.com/news/world-middle-east-11107739, accessed 11th June 2020.

Betts, A. and Collier, P. (2017). *Refuge: Rethinking Refugee Policy in a Changing World*. Oxford, UK: Oxford University Press.

Black, S., (2001). *Life and Debt*. United Kingdom: Axiom Films.

Corbin, J. (2008, June 10). BBC uncovers lost Iraqi billions. BBC. http://news.bbc.co.uk/2/hi/7444083.stm, accessed 11th June 2020.

FAO (2012). *Gender Inequalities in Rural Employment in Ghana: An Overview.* Gender, Equity and Rural Employment Division of Food and Agriculture Organization: Rome, Italy. http://www.fao.org/docrep/016/ap090e/ap090e00.pdf, accessed 11th June 2020.

FAO, IFAD, and WFP (2015). *The State of Food Insecurity in the World 2015: Meeting the 2015 International Hunger Targets: Taking Stock of Uneven Progress.* Food and Agriculture Organization of the United Nations: Rome, Italy. http://www.fao.org/3/a-i4646e. pdf, accessed 11th June 2020.

Finnegan, M. (2000, August 17). Cheney Gets $33.7-Million Retirement Deal, Firm Says. Los Angeles Times. https://www.latimes.com/archives/la-xpm-2000-aug-17-mn-5881-story.html, accessed 11th June 2020.

GCR (2018). The Global Compact on Refugees. A/73/12 (Part II). https://www.unhcr. org/5c658aed4.pdf, accessed 11th June 2020.

Glazebrook, T. (2011). Women and climate change: A case-study from northeast Ghana. *Hypatia*, 26(4), pp. 762–782.

Glazebrook, T. (2016). Gynocentric bio-logics: Anthropocenic abjectification and alternative knowledge traditions. *Telos*, 177, pp. 61–82.

Glazebrook, T. (2019). Are drone strikes ever ethical? *New Statesman* (4 November 2019). https://www.newstatesman.com/world/north-america/2019/11/are-drone-strikes-ever-ethical, accessed 11th June 2020.

Glazebrook, T. (2020). Abjection: The fate of the Banyamulenge in a world of failed humanitarian policy, unpublished paper scheduled at International Society for Military Ethics, Berlin, Germany, May 18–20, 2020 (Cancelled).

Glazebrook, T. and Story, M. (2012). The community obligations of Canadian oil companies: A case study of Talisman in the Sudan. In: Tench, R., Sun, W., and Jones, B., *Corporate Social Irresponsibility: A Challenging Concept*, Bingley, UK: Emerald Group Publishing, 231–261.

Griffin, S. (1971). Rape: The all-American crime. *Ramparts*, 10(3), pp. 26–35.

Hamada, K. (1974). An economic analysis of the duty-free zone. *Journal of International Economics*, 4(3), pp. 225–41.

Jensen, M. N. (2007, October 8). Newfound ancient African megadroughts may have driven evolution of humans and fish. *UANews*. https://uanews.arizona.edu/story/newfound-ancient-african-megadroughts-may-have-driven-evolution-humans-and-fish, accessed 11th June 2020.

Johnson, R.F. (2011). *Waging Peace in Sudan: The Inside Story of the Negotiations That Ended Africa's Longest Civil War.* Brighton, UK: Sussex Academic Press.

Kälin, W. (2012). From the nansen principles to the nansen initiative. *Forced Migration Review, 41* pp. 48–49.

Lasn, K. (1999). *Culture Jam.* William Morrow and Company, Inc.

Leopold, J. (2003, May 13). Defense Dept. secretly tapped Halliburton unit. *Global Policy Forum.* https://www.globalpolicy.org/component/content/article/185/40556.html, accessed 11th June 2020.

Long, J. (2013, March 28) Rape and sham marriages: The fears of Syria's women refugees. Channel 4 News. https://www.channel4.com/news/syria-women-rape-marriage-refugee-camp-jordan, accessed 11th June 2020.

MacKinnon, C. (1999). Rape: On coercion and consent. In: Pearsall, M., ed. *Women and Values*, 3rd ed. Belmont, CA: Wadsworth Publishing, 241–247.

Madani, D. (1999). *A Review of the Role and Impact of Export Processing Zones.* The World Bank, Development Research Group, Trade. http://documents.worldbank.org/curated/en/789981468766806342/pdf/multi-page.pdf, accessed 11th June 2020.

McAdam, J. (2016). Creating New norms on climate change, natural disasters and displacement: International developments 2010–2013. *Refuge, 29*(2), pp. 11–26.

Mies, M. and Shiva, V. (1993). *Ecofeminism*. London, UK: Zed Books.

Nansen Initiative (2015). *Disaster-Induced Cross-Border Displacement: Agenda for the Protection of Cross-Border Displaced Persons in the Context of Disasters and Climate Change, Vols. 1 and 2*. https://disasterdisplacement.org/the-platform/our-response, accessed 11ᵗʰ June 2020.

NPR (2007). *Millions Leave Home in Iraqi Refugee Service*. Scott Simon interviews Antonio Guterres. https://www.npr.org/templates/story/story.php?storyId=7466089, accessed 11ᵗʰ June 2020.

O'Donnell, K. and Newland, K. (2008). *The Iraqi Refugee Crisis: The Need for Action*. Migration Policy Institute.

OPEC (2020). *Brief History*. Organization of Oil Exporting Countries. https://www.opec.org/opec_web/en/about_us/24.htm, accessed 11ᵗʰ June 2020.

Pieraccini, F. (2020, January 8). The Deeper Story Behind the Assassination of Soleimani. Strategic Culture Foundation. World: Middle East. https://www.strategic-culture.org/news/2020/01/08/the-deeper-story-behind-the-assassination-of-soleimani/, accessed 11ᵗʰ June 2020.

PDD (2016). The Platform on Disaster Displacement. https://disasterdisplacement.org/, accessed 11ᵗʰ June 2020.

Pincus, W. and Milbank, D. (2004, June 17). Al qaeda-hussein link is dismissed. *Washington Post*. https://www.washingtonpost.com/wp-dyn/articles/A47812-2004Jun16.html, accessed 11ᵗʰ June 2020.

Roberts, M. (2009). War, climate change, and women. *Race, Poverty & the Environment, 16*(2), pp. 39–41.

Sharma, R. (2020). OPEC vs the US: Who controls oil prices? *Investopedia*. https://www.investopedia.com/articles/investing/081315/opec-vs-us-who-controls-oil-prices.asp, accessed 11ᵗʰ June 2020.

Statista (2020). *Jamaica: Unemployment rate from 1999 to 2019*. https://www.statista.com/statistics/527097/unemployment-rate-in-jamaica/, accessed 11ᵗʰ June 2020.

UDHR. (1948). *Universal Declaration of Human Rights* https://www.ohchr.org/EN/UDHR/Documents/UDHR_Translations/eng.pdf, accessed 11ᵗʰ June 2020.

UNFCCC. (2010). *Cancún Adaptation Framework*. https://unfccc.int/process/conferences/pastconferences/cancun-climate-change-conference-november-2010/statements-and-resources/Agreements, accessed 11ᵗʰ June 2020.

UNHCR (2010). *Text of the 1951 Convention on the Status of Refugees and Text of the 1967 Protocol Relating to the Status of Refugees*. https://cms.emergency.unhcr.org/documents/11982/55726/Convention+relating+to+the+Status+of+Refugees+%28signed+28+July+1951%2C+entered+into+force+22+April+1954%29+189+UNTS+150+and+Protocol+relating+to+the+Status+of+Refugees+%28signed+31+January+1967%2C+entered+into+force+4+October+167%29+606+UNTS+267/0bf3248a-cfa8-4a60-864d-65cdfece1d47, accessed 11ᵗʰ June 2020.

UNHCR. (2001–20). *Iraq Refugee Crisis*. https://www.unrefugees.org/emergencies/iraq/, accessed 11ᵗʰ June 2020.

UNSC (2008). *UN Security Council Resolution 1820 on women, peace and security*. https://www.unwomen.org/en/docs/2008/6/un-security-council-resolution-1820, accessed 11ᵗʰ June 2020.

Vatter, H. G. and Walker, J. F. (Eds.). (1995). *History of the U.S. Economy Since World War II*. Milton Park, UK: Routledge.

Waas, Murray (2005, November 22). *Key Bush Intelligence Briefing Kept Form Hill Panel*. National Journal Group Inc. http://www.leadingtowar.com/PDFsources_claims_collaboration/2001_09_21_NationalJournal.pdf, accessed 11th June 2020.

Walt, S. M. (2011, January 9). Wikileaks, April glaspie, and saddam hussein. *Foreign Policy*. https://foreignpolicy.com/2011/01/09/wikileaks-april-glaspie-and-saddam-hussein/#:~:text=In%20a%20now%20famous%20interview,or%20security%20commitments%20to%20Kuwait, accessed 11th June 2020.

Waring, M. (1988). *If Women Counted: A New Feminist Economics*. New York: Harper & Row.

Weerasinghe, S. (2018) *In Harm's Way*. UNHCR. https://www.unhcr.org/5c1ba88d4.pdf, accessed 11th June 2020.

World Bank. (1992). *Export Processing Zones*. Policy and Research Series, Vol. 20. http://documents.worldbank.org/curated/en/400411468766543358/pdf/multi-page.pdf, accessed 11th June 2020.

World Bank (2008). *Special Economic Zones*. http://documents.worldbank.org/curated/en/343901468330977533/pdf/458690WP0Box331s0April200801PUBLIC1.pdf, accessed 11th June 2020.

6

HUMAN GREED VERSUS HUMAN NEEDS

Decarbonization of the global economy

Furqan Asif

Introduction

From a human development perspective, people in the developed world, and increasingly the developing world, owe a lot of our modern lifestyle, technologies, and conveniences to the carbon emissions, spewed out of smoke stacks, tail pipes, etc. that are emblematic of 20th century industrialization and the economic system that has come to dominate the majority of the world. However, for all the benefits industrialization and capitalism have ushered in, many ills have manifested along with their evolution: sweeping and increasing economic inequality within countries (Knoop 2020), exploitative labour practices (Hannah and Peter 2014; Kaur 2010; Skrivankova 2010), creation of a new generation of elite capitalists (i.e. billionaires) (Petras 2008), and environmental degradation and destruction, to name a few. It is an inconvenient truth that the very systems our species (*Homo sapiens*) has invented have brought such rapid human progress yet are also now a threat to democratized human development, and to the future of humanity itself. Presently, there is no larger threat, not only to our survival but to non-human life as well, than climate change, for which our role in causing it is scientifically indisputable (Santer et al. 2019).

Like a scene from a murder mystery show, where the audience is asked to find who to blame for the crime, we can easily find the culprits responsible for the 'crime' of global warming and climate change: 90 companies are responsible for two-thirds of the world's carbon emissions from 1854–2010—half of this amount has been since 1984 and half have been emitted by the fossil fuel and cement production industry since 1986 (Heede 2014; Goldenberg 2013). When juxtaposed with the anticipated consequences of climate change impacts, such an analysis understandably warrant calls for decarbonizing our energy, transportation, and construction industries through increased adoption of renewable energy, deployment of low-carbon public transportation infrastructure, low-carbon building materials, and electric vehicles. At the same time, it must be asked: what ends have these carbon emissions been serving? At the core, the bulk of society's carbon emissions serve to power (both in a literal and figurative sense) our globalized, capitalist economic system. Factories and office buildings require a stable supply of energy to run. Workers need to fill their cars with gasoline so they can commute to work. Trains and buses that shuttle commuters from suburbs to cities require reliable supplies of diesel. In other words, our carbon emissions are intimately tied to the way our society has chosen to construct and organize its economic system, one which requires sustained and ever-increasing growth.

To avoid surpassing an increase of 1.5°C in global temperatures above pre-industrial levels, we have ten years (as of this writing) to ensure global carbon emissions decline by around 45%[1]. Moreover, by 2050, we must reach net zero emissions. This is the sobering conclusion of the Intergovernmental Panel on Climate Change (IPCC) *Special Report on Global Warming of 1.5°C* (IPCC 2018). The urgency and challenge are brought to the fore when you consider that some researchers conclude less than 2°C of warming by 2100 is unlikely (Raftery et al. 2017). Not to mention that analysis shows fossil fuel use has been *increasing*, not decreasing, and will continue to do so, particularly among the non-OECD countries (whose residents understandably desire a higher quality of life like their counterparts in OECD countries) (Never et al. 2020). Despite such daunting prospects, it is clear the challenge that human civilization faces, and what the IPCC states in its report unequivocally: we must decarbonize the global economy immediately to avoid the most catastrophic impacts from 'runaway' climate change. Despite the urgency and all the conferences, symposia, and committees formed over the last two decades (and more), meaningful progress has occurred at snail's pace. Current measures, while commendable, only inch us forward at a time when we need to *leap* forward.

There are two sides of the 'decarbonization coin' (so to speak). One is the technical, which includes fundamentally changing our energy sources, the materials we use in making things, and implementing techniques of carbon removal. We have a strong grasp of what is needed on a technical level, and indeed, many researchers note that all the technologies and strategies that we would need to deploy to reduce carbon emissions are already here and supplying 100% of our energy using hydroelectric, solar, and wind is achievable (Delucchi and Jacobson 2011a; Delucchi and Jacobson 2011b). Meanwhile, another group of researchers, those behind Project Drawdown, have developed a list of the most impactful strategies to implement and adopt to have the largest impact on carbon emissions, and all of these are things we can do today, right now (Makower 2017). There is no need to wait for nuclear fission or some of other miraculous technology to be invented and adopted. Having said that, while addressing climate change is in large part a technical problem (from a purely problem solving or engineering perspective), it exists within a globalized society, made up of 7.6 billion social actors. As such, climate change does not exist in a vacuum but instead is deeply interwoven and embedded within socio-economic and socio-political systems. This brings us to the other side of the decarbonization coin—the social—which includes societal norms, values, institutions, and culture (to name few). Coverage on climate change and decarbonization is largely made up of contributions by natural scientists and economists, and while they have made important contributions by investigating the intersection of the environment and the economy, the social dimension has seen less development and focus. If we have any chance of success at decarbonization, we must integrate social science perspectives within discussions around climate change and the drive to decarbonize, which acknowledge that we live our lives within structures of power (visible and not) and consider areas such as poverty, equity, (social) justice, inequality, gender, and social transformation.

As such, this chapter looks at the social side of the decarbonization coin through the adoption of the concept of basic human needs (Doyal and Gough 1984)—the 'human needs' in the title—and calls for a shift in framing the relationship between economy and society and its activities according to basic human needs. This is prefaced with a section of our current economic system which subsumes, and arguably compromises, human needs for the goal of increased profits and growth (the 'human greed'). The chapter starts with the latter which represents present conditions and then moves on to the former. The chapter concludes by suggesting a way forward by reimagining the relationship between the economy, the environment, and the social through principles focused on basic human needs that respect planetary boundaries.

Human greed

Over the past century, the prevailing paradigm shaping the architecture of the economic system has been neoclassical economics, and over the past few decades, this has transformed for many countries (e.g. the United States and the United Kingdom being key leaders) into a particular form—neoliberalism (Saad-Filho and Johnston 2005). Underlying these are certain key elements, under the banner of capitalism, broadly organized into two general areas: production and consumption. Historically, these elements have functioned on a microlevel in villages and communities, evolving into forms such as bartering (i.e. offering goods or services in exchange for those needed by the other person). The widespread adoption of money and currencies, elements central to the capitalist system, reshaped the model around production and consumption because people were no longer exchanging goods and services based on 'use-value' (i.e. the material uses of the object and the human needs it fulfills) but 'exchange value' (i.e. two commodities are exchanged on the 'open market' by being compared to a third item that acts as a 'universal equivalent'—money). In other words, one of the outcomes of capitalism is that it puts primacy on exchange value, and the drive to accumulate it, and subordinates use-values, i.e. human needs. Fundamentally, the most crucial element is the *private* ownership of the means of production and its operation for profit. This results in private firms controlling the means of production, not those producing it, and being owners of the enterprise (the 'capitalists'). Another element that is important for a 'free market' is well-defined property rights.

However, there are activities related to the means of production whose property rights are not well-defined, collectively called externalities in the economics parlance, which are costs involved in production but which are not borne by the firms doing the production, but instead third parties (typically without their consent). In the context of this chapter, the most relevant (negative) externality is carbon emissions (and related to this, pollution from burning of fossil fuels more generally). Historically, economics has tended to do a poor job of factoring in these negative externalities within their models or analysis, and some firms chalked them up (crudely) as 'the cost of doing business'. Yet, climate change and its ensuing impacts in the coming decades makes such a perspective outdated, out of touch with reality, and arguably, irresponsible. In short, externalities matter and fossil-fuel related emissions (carbon, methane, nitric oxides, sulphur, etc.) are some of the most egregious. When you consider that more than 70 percent of global emissions come from just 100 companies, those at the helm, the CEOs, have a disproportionate influence on such negative externalities as they plan the activities of their companies. Indeed, there is an argument to be made that billionaires, through their position as leaders of these companies are complicit in causing and exacerbating climate change (Darby 2018).

For the past century, fossil fuel companies have been getting away with not being responsible for the negative externalities their activities created. The experience of adopting market-based mechanisms to encourage responsibility and to address such externalities do not encourage much faith in free-market principles saving us from climate catastrophe. Recent analysis has shown even the most aggressive carbon tariffs (or 'taxes'), i.e. $100 USD per ton of CO_2 will have limited impacts on the fossil fuel industry (Barron 2018). Besides such conclusions, we can also see proof of the inadequacy of market-based instruments to facilitate decarbonization by considering that increasing adoption of carbon tariffs across countries has had modest impact on carbon emissions. Moreover, energy companies have little incentive to shift from a business as usual scenario because to do so is expensive, at least in the short term, which is the time window by which most corporations are assessed: profit and revenue per quarter, growth per year, CEO's yearly bonuses. Even on a longer time horizon of decades, fossil fuel companies' complacency is emboldened by trends that show fossil fuel demand and use will only *increase* due to demand rapidly urbanizing Global South, as demand tapers or reduces in advanced economies (IEA 2020). For example, world energy consumption is expected to rise by 50% between 2018 and 2050, with almost all of this growth expected in non-OECD countries, and with Asia accounting for most of the increase in energy use (EIA 2019).

Human needs

The concept of human needs as put forward by Doyal and Gough (1984) defines *basic individual* (human) *needs* as elementary goals that have to be achieved before an individual can achieve any other goal, regardless of their culture, creed, religion, etc. As the authors describe, basic human needs are made up of two categories, survival/health and autonomy/learning. Survival entails all individuals having of their physical requirements being met, e.g. food, water, while also having sense of their identity and autonomy to carry out actions, i.e. autonomy or agency. Thus, both survival and autonomy are necessary to carry out the achievement of other goals. Moreover, health, both physical and mental health, constitute the most basic human need, and one which must be satisfied before any others (Doyal and Gough 1984).

Outside of the individual, and in recognizing the importance of socialization to our species, there are also basic societal needs that must be achieved so that individuals can satisfy their basic needs: production, reproduction, culture/communication, and political authority (Doyal and Gough 1984). Indeed, individual needs and societal needs are interrelated and depend one another. Crucially, this framework of basic needs is prefaced by the need for human *liberation*, i.e. the goal of maximizing individual or collective *choice* in meeting basic needs. Arguably, our current economic system of capitalism does not lend a large portion of the human population much choice in how they meet their basic needs. For example, a woman from a poor household in rural Lesotho is compelled to leave the home and work in a garment factory to support her family's meagre income so that their basic human needs can be met (Baylies and Wright 1993). In other instances, poor small-scale fishers are 'forced' to continue fishing despite declining incomes because of limited alternative options to support their livelihoods (Asif 2020).

At the same time, there are defined ecological boundaries that place a limit on economic production which must be reconciled with the need to have sufficient material resources to support the satisfaction of basic human needs for all. From this perspective, there is an impetus to employ strategies that protect and conserve the environment given our reliance on the natural earth systems for our survival. In other words, it is not only a matter of natural resources being available, but a question of how and why we use them. From this, the concept of 'ecological needs' (Doyal and Gough 1984) helps to explain the oppositional forces of human greed and human need, which considers the global, natural constraints within which all actions to achieve basic individual needs must function.

The logic embedded within free market principles and capitalism requires ever-increasing economic growth. What does this mean on a practical, everyday level? Such a system is fundamentally antithetical to providing for basic human needs because the profit motive supplants everything else. To most individuals who are deeply fearful of the consequences of a warmer world and what this means for their children's and grandchildren's future, such trends do not lend an optimistic perspective. We must shift our focus within decarbonization's goal to centre on expanding the possibilities for everyone to achieve their basic human needs.

In reconciling human needs and human greed, it becomes clear that there will have to be significant and profound constraints placed on vested interest groups that are furthering human greed at the cost of human needs. The following section outlines where things stand, the challenges we face, and possibilities of a more sustainable future.

Discussion

The climate change space has been imbued with the narrative portraying climate change as a foe that must be fought 'against' in a kind of battle that we must win. This can be seen quite literally, in fact, with phrases such as 'the fight against climate change' (see Anderson and Nevins 2016; Bortscheller 2009; Hickey, Rieder and Earl 2016; Hunt 2009). However, once we start thinking about the social side of the decarbonization coin, it forces us to refocus the question from 'what are we fighting against?' to 'what are we fighting *for*?'. Thinking from the social side of decarbonization demands us to think and act differently, and most importantly, it requires us to change our value system—from one where our institutions are not driven solely by economic growth and profit-earning and where we are not driven by (and to) (over) consumption.

The essence of this is summed up well by Weizsäcker and Wijkman (2017), the two co-presidents of the Club of Rome (the same group that published in 1972 the controversial, albeit ground breaking, *The Limits to Growth* which warned of global collapse from a business as usual pathway) in their recent volume on the intersection of the economy and a sustainable future:

> "Values represent the quintessence of human wisdom acquired over centuries. And in the new system that's developing, they must embody the fundamental principles for sustainable accomplishment, whether individual or social. These must be even more than the inspiring ideals that supply the energy needed to fulfil human aspirations. Values are a form of knowledge and a powerful determinant of human evolution. They are psychological skills that have profound *practical* importance. Education must be founded on values that promote sustainability and general well-being for all. A move toward inculcating sustainable values would amount to a paradigm change in our current society's value system. It would consider as its aim the greater well-being of both human and the natural systems on which they depend, rather than a valuation for more production and consumption. Conscious emphasis will be placed on values that are truly universal, as well as on respect for cultural differences. At the grass-roots level, the movement towards sustainability can build on deep local values. Values can create transformational leadership, leadership in thought that leads to action." (Weizsäcker and Wijkman 2017, p. 198)

A fundamental challenge in the context of climate change and a kind of 'catch-22' of achieving decarbonization is that values do not change overnight, they take time—something that we do not have an abundance of given our ten year window. Adding to this, one of the consequences of 'human greed'—the continued push to profit off of fossil fuels as long as possible—that will pose an exceptional challenge to decarbonization of the global economy is 'carbon lock-in'. The term relates to the nature of fossil-fuel infrastructure which generally is designed to last several decades. Thus, as countries build such infrastructure, they are in a sense guaranteeing or are 'locked' to emitting a certain amount carbon for the medium-term. As Seto et al. (2016) explain, there are three main types of carbon lock-in that will stifle progress towards decarbonization: i) infrastructural and technological lock-in—e.g. street layouts, land use patterns, buildings, long-lived fossil fuel distribution and consumption infrastructure; ii) governance, institutional, and decision-making associated with energy production and consumption, which influences energy supply and demand; and iii) behaviour, habits, and norms related to demand for energy-related services. For industrialized countries, their long-standing reliance on fossil fuels to provide energy presents an opportunity to transition away from carbon lock-in because as their energy infrastructure ages, there is a window to adopt sustainable energy systems. The larger challenge in addressing carbon lock-in will be for industrializing countries who are rapidly building infrastructure that rely on fossil fuels instead of systems that employ renewable energy sources.

Dismayingly, the development of carbon-intensive infrastructure is *increasing*, not decreasing. This boom in fossil fuel infrastructure creation has been facilitated by the passing of key pieces of legislation which loosens regulations on the fossil fuel industry. For example, in the United States, the Energy Policy Act of 2005 not only provided tax incentives and loan guarantees to companies for energy production but also exempted the fluids used for natural gas production and hydraulic fracking from protections under the Clean Air Act, Clean Water Act, Safe Drinking Water Act, and the Comprehensive Environmental Response, Compensation, and Liability Act[2]. Combined with the lifting of the 40-year ban on exporting domestic fossil fuel production, this has all but guaranteed carbon lock-in and hampers the path to decarbonization. It is important to point out here that much of this boom in carbon-intensive infrastructure and production is designed for export, primarily to be used in plastics production—not energy. Since 2010, fossil fuel companies have spent $180 billion on new 'cracking' facilities (cracking is a petrochemistry technique to break down crude oil into various useable components, including petrochemicals for plastics) (Taylor 2017). The production of petrochemicals is expected to increase from 16% of oil demand in 2020 to 20% by 2040, and a sizeable amount of this is to supply the raw materials needed to make plastics (McKay 2019). Such trends will result in further carbon lock-in for the next several decades, which is the expected lifespan of the kinds of infrastructure being built.

Amidst these observations, something must change for decarbonization of the global economy to become a possibility. Unfortunately, we cannot simply expect change from our governments, we must *demand* it. After all, if history is to be used as a harbinger, many achievements and milestones in social justice (because, to be sure, climate change is as much, if not even more, of a social justice issue than a technical issue) and progress were achieved directly because of grassroots, citizen action-led movements. For example, the women's suffrage movement which paved the way for women's rights to vote in the mid-19th century. National and international organizations, namely the International Women Suffrage Alliance, were formed to coordinate efforts around the world to achieve this objective (Sneider 2010). These organizations developed extensive political campaigns which were crucial in obtaining the necessary legislation and constitutional amendments to achieve women's suffrage. In other cases, progress was made despite special interest (read: profit maximizing) organizations resisting change and because we collectively decided that there must be change. Child labour was commonplace in coal mines in the 17th to 18th century during the industrial revolution with children as young as four years old being put to work in dangerous conditions for pennies a day. The public was largely unaware of this until a serious accident (thunderstorms led to a stream overflowing into a ventilation shaft, killing 26 children) revealed the extent of child labour used in the mines (Simkin 2020). Public pressure alongside pressure from Queen Victoria led to establishing of the Royal Commission of Inquiry into Children's Employment and the release of a report which caused widespread public outrage. Despite this, three-quarters of petitions to Parliament were against the proposed regulation outlawing child labour. Unsurprisingly, upwards of 86 percent of these petitions came from districts with a high concentration of child labour and where employers feared that this new law would lead to less profits (Kirby 2003). This historical example reiterates the point made earlier, namely that capitalism and the lust for profits by individuals and organizations obfuscates the importance of meeting basic human needs.

Conclusion

Some encouraging news is that achieving energy solely from water, wind, and the sun is possible and analysis by Delucchi and Jacobson (2011b) shows that the cost of doing so is similar to the current cost of generating energy from fossil fuel (using figures from 2007 in the United States). The authors emphasize, and reiterating the point made in this chapter, that the barriers to do so are largely social and political, not technological, or economic. Perhaps the most inspiring case of humans working together in solidarity to prevent an environmental catastrophe is the signing of the Montreal Protocol in 1987 and its ratification by 196 countries. Officially known as the Montreal Protocol on Substances That Deplete the Ozone Layer, it paved the way for phasing out of ozone-depleting substances, known as chlorofluorocarbons (CFCs) and hydrochlorofluorocarbons (HCFCs) from industrial use (e.g. refrigerants), that scientists discovered were breaking down the ozone layer in the upper atmosphere, a critical invisible sheath that protects earth from the most harmful ultraviolet radiation. In doing so, 1.5 million cases of skin cancer, 330,000 cancer deaths, and 129 million cases of cataracts were prevented, by some estimates (Cardoni 2010). Moreover, the Montreal Protocol also played a key role in climate change by preventing 11 billion tons of CO_2 from entering the atmosphere (the CFCs and HCFCs act as even more powerful heat-trapping molecules than CO_2).

In taking stock of some past achievements, we must ask ourselves: which way should we reengineer our economic and social systems to meet the basic human needs of 7.6 billion humans? As I put forward in this chapter, the primacy of ensuring basic humans needs should be the underlying principle that guides all future policies within governments and organizations if we want to have any chance of not putting a permanent burden of coping with the impacts of climate change onto the future generations of humans, all of whom would have played no part leading into their existence. Going forward, we must remind ourselves of our past and those specific instances where we overcame pressing social and environmental challenges. The way this was achieved is that we united, we cared, we did more than just vote every four years—we formed coalitions, cooperatives, encouraged activism, raised awareness, and demanded change from our institutions and politicians, who serve us. We have done it before. And we can do it again.

Notes

1 It must be noted that an important requirement to achieve this involves not only reducing carbon emissions but also *removing* carbon dioxide currently in the atmosphere because of the time lag of carbon emissions, i.e. the climate impacts of carbon dioxide emitted takes some time (about a decade on average) to manifest given the way the global climate and various earth systems function, which increases with the size of emissions (see Zickfeld and Herrington 2015)

2 The Energy Policy Act also contains more environmentally progressive measures such as tax credits for wind and other alternative energy producers, tax breaks for those making energy conservation improvements to their homes, protects the Great Lakes from fossil fuel exploration, among others.

References

Anderson, K. and Nevins, J. (2016). Planting seeds so something bigger might emerge: The Paris agreement and the fight against climate change. *Socialism and Democracy*, 30(2), pp. 209–218.

Asif, F. (2020). *Coastal Cambodians on the Move: The Interplay of Migration, Social Wellbeing and Resilience In Three Fishing Communities* [unpublished]. Thesis, Université d'Ottawa/ University of Ottawa. Available from: http://ruor.uottawa.ca/handle/10393/40420 [accessed 24 April 2020].

Barron, A.R. (2018). Time to refine key climate policy models. *Nature Climate Change*, 8(5), pp. 350–352.

Baylies, C. and Wright, C. (1993). Female labour in the textile and clothing industry of Lesotho. *African Affairs*, 92(369), pp. 577–591.

Bortscheller, M.J. (2009). Equitable but ineffective: How the principle of Common but differentiated responsibilities hobbles the global fight against climate change. *Sustainable Development Law & Policy*, 10 p. 49.

Cardoni, S. (2010). Top 5 Pieces of Environmental Legislation. *ABC News* [online], 1 July 2010. Available from: https://abcnews.go.com/Technology/top-pieces-environmental-legislation/story?id=11067662 [accessed 9 August 2020].

Darby, L. (2018). *Billionaires Are the Leading Cause of Climate Change* [online]. *GQ* [online]. Available from: https://www.gq.com/story/billionaires-climate-change [accessed 16 October 2018].

Delucchi, M.A. and Jacobson, M.Z. (2011a). Providing all global energy with wind, water, and solar power, part I: Technologies, energy resources, quantities and areas of infrastructure, and materials. *Energy Policy*, 39(3), pp. 1154–1169.

Delucchi, M.A. and Jacobson, M.Z. (2011b). Providing all global energy with wind, water, and solar power, part II: Reliability, system and transmission costs, and policies. *Energy Policy*, 39(3), pp. 1170–1190.

Doyal, L. and Gough, I. (1984). A theory of human needs. *Critical Social Policy*, 4(10), pp. 6–38.

EIA. (2019). *International Energy Outlook 2019*. U.S. Energy Information Administration (EIA).

Goldenberg, S. (2013). Just 90 companies caused two-thirds of man-made global warming emissions. *The Guardian* [online], 20 November 2013. Available from: https://www.theguardian.com/environment/2013/nov/20/90-companies-man-made-global-warming-emissions-climate-change [accessed 13 August 2020].

Hannah, L. and Peter, D. (2014). *Precarious Lives: Forced Labour, Exploitation and Asylum*. Policy Press.

Heede, R. (2014). Tracing anthropogenic carbon dioxide and methane emissions to fossil fuel and cement producers, 1854–2010. *Climatic Change*, 122(1), pp. 229–241.

Hickey, C., Rieder, T.N. and Earl, J. (2016). Population engineering and the fight against climate change. *Social Theory and Practice*, 42(4), pp. 845–870.

Hunt, C.A.G. (2009). *Carbon Sinks and Climate Change: Forests in the Fight Against Global Warming*. Edward Elgar Publishing.

IEA. (2020). Global CO2 emissions in 2019—Analysis. *International Energy Association (IEA)* [online], 11 February 2020. Available from: https://www.iea.org/articles/global-co2-emissions-in-2019 [accessed 17 August 2020].

IPCC. (2018). *IPCC Special Report - Global Warming of 1.5°C: Summary for Policymakers* [online]. Intergovernmental Panel on Climate Change. Available from: http://report.ipcc.ch/sr15/pdf/sr15_spm_final.pdf [accessed 23 October 2018].

Kaur, A. (2010). Labour Migration in Southeast Asia: Migration policies, labour exploitation and regulation. *Journal of the Asia Pacific Economy*, 15(1), pp. 6–19.

Kirby, P. (2003). *Child labour in Britain, 1750–1870*. 2003 edition. New York: Palgrave.

Knoop, T.A. (2020). *Understanding Economic Inequality: Bigger Pies and Just Deserts*. Edward Elgar Publishing.

Makower, J. (2017). 'Drawdown' and global warming's hopeful new math. *GreenBiz* [online], 17 April 2017. Available from: https://www.greenbiz.com/article/drawdown-and-global-warmings-hopeful-new-math [accessed 21 October 2017].

McKay, D. (2019). Fossil fuel industry sees the future in hard-to-recycle plastic. *The Conversation* [online], 10 October 2019. Available from: http://theconversation.com/fossil-fuel-industry-sees-the-future-in-hard-to-recycle-plastic-123631 [accessed 17 August 2020].

Never, B., Albert, J.R., Fuhrmann, H., Gsell, S., Jaramillo, M., Kuhn, S. and Senadza, B. (2020). Carbon consumption patterns of emerging middle classes. *Discussion Paper* [online], 2020. Available from: https://www.die-gdi.de/discussion-paper/article/carbon-consumption-patterns-of-emerging-middle-classes/ [accessed 8 May 2020].

Petras, J. (2008). Global ruling class: Billionaires and how they 'make it'. *Journal of Contemporary Asia*, 38(2), pp. 319–329.

Raftery, A.E., Zimmer, A., Frierson, D.M.W., Startz, R. and Liu, P. (2017). Less than 2°C warming by 2100 unlikely. *Nature Climate Change*, 7, p. 637.

Saad-Filho, A and Johnston, D. (2005). *Neoliberalism: A Critical Reader* [online]. University of Chicago Press. Available from: https://econpapers.repec.org/bookchap/ucpbkecon/9780745322995.htm [accessed 17 August 2020].

Santer, B.D., Bonfils, C.J.W., Fu, Q., Fyfe, J.C., Hegerl, G.C., Mears, C., Painter, J.F., Po-Chedley, S., Wentz, F.J., Zelinka, M.D. and Zou, C.-Z. (2019). Celebrating the anniversary of three key events in climate change science. *Nature Climate Change*, 9(3), pp. 180–182.

Seto, K.C., Davis, S.J., Mitchell, R.B., Stokes, E.C., Unruh, G. and Ürge-Vorsatz, D. (2016). Carbon lock-in: Types, causes, and policy implications. *Annual Review of Environment and Resources*, 41(1), pp. 425–452.

Simkin, J. (2020). *Child labour in the mining industry* [online]. *Spartacus Educational* [online]. Available from: https://spartacus-educational.com/Child_Labour.htm [accessed 13 August 2020].

Skrivankova, K. (2010). Between decent work and forced labour: Examining the continuum of exploitation, 2010, p.38.

Sneider, A. (2010). The new suffrage history: Voting rights in International Perspective'. *History Compass*, 8(7), pp. 692–703.

Taylor, M. (2017). $180bn investment in plastic factories feeds global packaging binge. *The Guardian* [online], 26 December 2017. Available from: https://www.theguardian.com/environment/2017/dec/26/180bn-investment-in-plastic-factories-feeds-global-packaging-binge [accessed 16 August 2020].

Weizsäcker, E.U. von and Wijkman, A. (2017). *Come On!: Capitalism, Short-Termism, Population and the Destruction of the Planet*. Springer.

Zickfeld, K. and Herrington, T. (2015). The time lag between a carbon dioxide emission and maximum warming increases with the size of the emission. *Environmental Research Letters*, 10(3), p. 031001.

7

THE FALL OF THE DOLLAR

Abdullah Al Mamun and Sanni Yaya

Introduction

The fall of the dollar has been predicted for several decades. Scholars, investors, market analysts and so on forecasted the demise of the dollar. For instance, Robert Triffin argued in the nineteen sixties that the dollar would lose its supreme position among all the currencies (Triffin 1960). Similarly, just after a decade, economic historian Charles Kindleberger claims that the power of greenback as an international currency is over (Kindleberger 1976). Continuing in the same vein Canadian billionaire investor Ned Goodman delivered a speech in an investment conference in 2013 that the era of the US Dollar as the world's reserve currency is over (The Northern Miner 2013). Even today, these dire predictions on the declining date of the dollar continue unabated in both print and electronic media. These prognostications are closely followed by predictions on what would happen in the aftermath (The Balance 2020).

However, scholars and analysts are also very much optimistic about the good future of dollars. For example, Lee argued that the US dollar will remain the world's main currency of choice (Lee 2019). Besides, Richter reports that the efforts to diminish the dollar's dominance will not work anytime soon (Richter 2018). Moreover, Best claims that the dollar is the topmost foreign reserve currency in the world (Best 2020).

In this chapter we analyze the status dollar as the most important international currency. It will also explore some of the core characteristics of currency internationalization and will examine the merit of currency competitors, such as the euro and the renminbi (RMB or yuan) against the dollar in the light of some core financial and political factors. Furthermore, it will offer a prospect of Chinese digital currency as a potential actor in the world market in brief.

The dollar centrality: How did that happen?

What was known as the "international gold standard" until the early 1930s was the British "pound sterling" as the international currency. In reality, Great Britain's printed paper money or simply Bank of England's credit that served as the international currency. The pound sterling was holding this position was because of the strength of the British Empire's economic, military and other strengths. British money or pound sterling rose to that prominence in the late 19th century and that hegemony, (i.e., leadership) remained in place until the early 1930s when the central bank of UK was ready to give a definite amount of gold (at a fixed rate) to anyone who would like to cash their sterling's designated in the paper (Lindert 1969). That was the secret of the gold standard. Pound sterling freely exchangeable against gold gave meaning to the phrase "as good as gold." As long as pound sterling was in demand as a currency by foreign nations for trade and investment, the rule of pound sterling remained intact, however, the US dollar or greenback had taken the place of sterling gradually over the time.

The greenback started to replace sterling as the leading reserve currency in the mid-1920s. The status of dominant currency once lost for sterling but was not lost forever. In the early 1930s, sterling surpassed the dollar as the leading currency. It makes the fact clear that reserve-currency status depends not only just economic, commercial and financial size but also on the issuer's political influence. The inter-world war period the witness of the rise of New York as a financial centre, rivalling London. This eventually and made the dollar as an international currency, rivalling sterling. The question exactly when the dollar overtook sterling as the leading currency in which to hold foreign exchange reserves. The answer sheds light on the competitive history for leading financial status between London and New York.

Eichengreen and Flandreau report that the dollar began to surpass the sterling as a leading reserve currency in the 1920s. With the post-world war crisis, the pound sterling lost its hegemony and eventually fell in 1931 during the great depression. It was triggered by the fall of stock prices in 1929 in the New York Stock Exchange. They also argued that the global foreign exchange reserves were on the order of more than two billion dollars which was a significant amount on that time as a reserve currency (Eichengreen and Flandreau, 2008). On the contrary, Triffin states that sterling occupied the leading position as a reserve currency in the 1920s and 1930s despite a declining trend. He estimated that the sterling's share as a reserve currency was 80 percent in 1928, and in the next ten years, the share decreased by ten percent (Triffin 1960). Besides, Aliber claims that the sterling was in the position of number one as an international currency until the Second World War (Aliber 2002). Furthermore, Chinn and Frankel argue that the rise of the dollar began after the Second World War and remained in place until 1971 when US president Richard Nixon decided to end the dollar's gold convertibility in 1971 (Chinn and Frankel 2008). However, despite the termination of the dollar's gold convertibility, the dollar remains in the race of the most important international currency along with the euro and the renminbi.

Currency power: An overview of monetary rivalry

The prediction about the decline of greenback's primacy and supremacy is not recent. One of the reasons for the dollar pessimism would be – for the first time, the world's greatest monetary power is also the world's biggest debtor indicating the accumulated debt by the United States. Eichengreen argues that increasing, "doubts are pervasive about whether the dollar will retain its international role" (Eichengreen 2011 p. 121) The earlier opinion of Benjamin Cohen was that "The world is approaching towards a leaderless currency system," with several currencies in contention and none clearly in the lead. (Cohen 2009).

The world is rapidly moving away from the solo primacy of the dollar. Many economists hold the opinion that both the euro and the RMB have the capability to challenge the dominance of the greenback (Eichengreen 2011). Rather, the world market is most likely to be dominated by three currencies. According to the World Bank, the most likely scenario for the international monetary system is a multicurrency system centered around the U.S. dollar, the euro, and the renminbi (World Bank 2011).

The old debate "The Euro vs Dollar" is not unreasonable and unjustified. The promising appearance of the Euro at its birth convinced many observers that it can challenge the reign of the dollar as an international currency. The large economic base, political stability, low inflation rate, etc. indicated to Euro's capability to fulfill the necessary attributes to become the supreme international currency.

But the 2008 crisis pointed out serious structural flaws of the Euro. Especially, it brought out the question, among the 28 member nations "who is in charge of this currency?" (The Economist, 2013). Several scholars argue that the euro rivalling the dollar is gone. And the failure of the euro to compete with the greenback's primacy made many turn their focus on the RMB as the new century belongs to China.

The economist Arvind Subramanian predicts with confidence that, China's growing size and economic dominance are likely to translate into currency dominance. The renminbi could surpass the dollar as the premier reserve currency well before the middle of the next decade. (Subramanian 2011)

The dollar optimism, on the other hand, is also standing on strong foundations. The US financial sector is controlling the global network of currencies and capital markets. As Rajendran writes, "America has come to function as a sort of central processing core through which funds are routed" (Rajendran, 2013).

The debate around the position of the leading international currency is an ongoing topic for scholars, analysts and so on. Several scholars argue that the leadership of a currency has a strong relationship with meeting the prerequisites of currency internalization.

Prerequisites for an international currency

The internationalization of a currency depends on various factors. According to Cohen, There are five basic prerequisites, financial development, economic size, military reach, foreign policy ties, and effective governance, which are needed to have success for a country to make its currency as international currency (Cohen 2015). These five prerequisites together form a combination of both economic and political factors. A country has to have a good measure of control over these factors to sustain a good appeal in the world market about its currency.

Economic size: The first pre-requisite is the relative size of the economy or the relevant nation's share of the global product. The economic size is measured by the gross domestic product (GDP) of a country and by its importance in world trade. Frankel argues that the country that has the largest economic size in the world, gets a natural advantage in the competition of currency internationalization (Frankel 1995). A larger economy of a country in the world builds a greater weight for its currency in international commerce compare to a currency issued by any country. Besides, according to history, most of the international currencies were backed by a powerful economy. For example, the sterling was one of the most demandable international currency because of larger economic size in the world (Dawnay 2001). A big and stable economy establishes a network value of the currency in the world market. Bayoumi and Eichengreen also state that a bigger economic size would build trust in its consumers as a stabilized currency (Bayoumi and Eichengreen 1998). Besides, Bacchetta and van Wincoop argue that the economic size of a currency issuing country plays a significant role in international trade (Bacchetta and van Wincoop 2005). Moreover, Engel report that both local firms of a country that has a large size economy get enough freedom to invoice their exports and imports in local currency. He also argued that not only the local firms but also the foreign firms use the same currency to invoice some, if not all of their trade to get financial advantages (Engel 2005).

Financial development: Financial development indicates sophisticated banking and open capital markets of a currency issuer's country. Cohen report that "To appeal to outside investors or central banks, a currency must offer the qualities of exchange convenience and capital certainty – a high degree of transactional liquidity and reasonable predictability of asset value" (Cohen 2015, p. 9). He also argues that the financial market of a currency-issuing country has to offer three basic essential characteristics. Firstly, the capacity to tolerate relatively large market orders without impacting expressively on an individual asset's price. Secondly, broad trading volumes and sufficient market competition to guarantee that the spread between ask (sell) and bid (buy) prices is trivial. And thirdly, the capability of market prices to recuperate quickly from the orders of unusual buy and sell orders (Cohen 2015). Besides, Kenen report that sophisticated baking and open capital markets allow foreign official institutions to exercise the option of holding the country's currency and financial instruments on a significant scale (Kenen 2011).

Foreign policy ties: Foreign policy ties of a country with other countries play a very important role in influencing currency preference in trade and commerce. Also, a leadership position of a country in a region or on the world would significantly increase the use of the issuer's currency by its partner countries. Li argues that the geopolitical hegemony of a country earns the trust of other governments in its currency. The tighter the relationship between the country at the leadership position and its allies, the greater is the encouragement they feel to use the hegemon's currency. Even formal alliances have a distinct impact on a country's choice of an anchor currency, as frequently found in many empirical studies (Li 2003). Besides, Posen claims that foreign policy failure would lead to displacements from a leading role. For example, he reported, "some of the declines in the dollar's global role of late is already due to the foreign policy failures of the Bush administration, not just to current account imbalances and financial turmoil." (Posen 2008, p. 1).

Military outreach: A country or nation with powerful militarily can provide a safe environment for middling or even long-term investments. It creates great appeal for cautious foreign investors and banks. Strong military power would offer security leadership in any region and would negotiate with the regional countries to use the currency. This process assists in strengthening the currency as an international currency. Chey states that strong military power aided the UK to strengthen the sterling in the postwar period, and the dollar in the 1960s (Chey 2012).

Effective governance: The scope of effective governance is vast for political scientists. But in this issue the broad concept is framed into a successful and sustainable policy that can maintain a relatively low level of inflation and inflation variability over time. A currency must provide a good degree of security about its purchasing power if its world-wide use is expected to be ensured (Tavlas and Ozeki 1992). Moreover, political stability through the rule of law, effective law about property rights are indicators of effective governance regarding this issue (Lim 2006). Sobel reports that effective governance was necessary to have success in currency internationalization for all the major currency issuer countries. This issue is also true in the present era including the dollar, sterling, Deutsche mark, Japan's yen, and the euro (Sobel 2012).

Considering these finance, trade and policy prerequisites, the following sections examine the three main competitors in the race of the most deserving international currency.

Status of the dollar, euro and renminbi in the financial and political sectors

The dollar

US debt is one of the major issues to the dollar pessimists. Its size is growing bigger over time and became the world's largest debtor. According to the US Debt Clock, the amount of US national debt on July 8, 2020, is more than 26 trillion dollars (Usdebtclock 2020). It might be possible that the dollar would lose its position anytime if the United States does not take necessary actions (Wheatley 2013).

However, it is hard to find some strong evidence of policies to control consumer behaviour of the United States in the last two decades. Calleo argues that the Americans are habituated to their "exorbitant postwar privileges." They prefer to export more dollars rather than consume less (Calleo 2009).

Despite of enormous debt, the dollar is attracting the attention of investors. Also, the US dollar is holding the primary position of the foreign exchange reserve for a long time. According to the IMF, the US dollar holds approximately 62% of foreign currency reserves in the world (IMF 2020). Cohen claims that the dollar's exclusive advantages as a transnational store of value. The United States provides a great set of financial markets, promising liquidity and safety as well as broad network externalities in trade, political ties, massive military reach, and efficient democracy (Cohen 2015).

The US financial sector is one of the main attractions for investors in the world. It offers a great range of financial instruments, such as stocks and bonds, swaps, options, forward and futures contracts, derivatives and so on. It is also open to all. There is a risk, but the reward is also high for the investors. Though some other currency issuer countries also provide almost the same financial instruments, the scale of trading is not is as large as the United States. Rajendran reports that the United States became the core funds processing center in the world (Rajendran 2013). Besides, Oatley claims that most of the countries in the world have direct strong financial ties with the United States that makes the dollar in superior position compare to any currency in the world (Oatley et al. 2013). Moreover, Fields and Vernengo state that the most important feature of a dominant currency is that it offers a secure financial asset that smooths the functioning of financial markets and dollar is ahead among all the currencies in offering that feature (Fields and Vernengo 2013).

The United States is also in the leading position in trade externalities in the world. It is now the second-largest exporting country, just after China (Duffin, 2020). At the same time, US companies dominate more than 70% of the world economy (Starrs 2013). According to Tett, the United States is the primary source of supplying safe liquid assets in the field of trade and commerce at a greater scale in the world (Tett 2014).

Regarding the political factors, the United States has a great deal of political control in the world. The United States has approximately 800 bases in more than 70 countries and territories (Vine 2015). Though there are some incidents where Russia and China showed strong diplomacy over the United States in recent times. For example, Kempe argues that Chinese and Russian involvement in Syria and Afghanistan shaken the US's global leadership (Kempe 2019). However, these two countries are suffering from some built-in disadvantages, such as aging demographics, slowing growth, Hong Kong protests, and authoritarian structures that would hinder them to be in the global political leadership in the long run (Kempe 2019). Grey also claims that despite some ups and downs, the United States plays a leading leadership role everywhere in the globe with a few exceptions (Grey 2019). This political leadership of the United States helps the dollar to dominate the other currencies in the monetary rivalry.

The euro

The euro promised competitive success and challenged the primacy of the greenback as the international currency. The European Central Bank, one of the world's most powerful monetary authorities ensured the currency's stable and

vast economic base, unique political stability and an obvious low rate of inflation. The value and cross-border usability of the Euro seemed more than secure.
During the time of the Euro's introduction in the market, Mundell reports that
the newfound currency would contest the dollar and would alter the power con-
figuration of the system (Mundell 2000).

But, in 2008, the great crisis of the world economy struck hard all the optimistic calculations and predictions about the Euro. Portugal, Spain, Greece and
Cyprus with their sovereign debt problems have shaken confidence about Euro.
Right after birth internationalization of euro was running very swiftly. But by
the mid-2000s, a changed situation appeared as the cross-border use of Euro had
already levelled off. And it has even considerably slipped back in most recent
times. For instance, The Economist reports that hot talk of the euro rivalling the
dollar is gone (The Economist 2013).

The query would be, what went wrong with Euro and how its strong challenge
to dollar faded eventually? First, the euro is a currency without a country. Its members are a club of sovereign states. Unlike the dollar which is created and managed
by a single sovereign power, Europe's money is the product of a multi-state agreement – the Maastricht Treaty, EMU's founding document. Decision making about
monetary management is not smooth here because of conflict between the domain
of the Euro-zone and the legal jurisdictions of its participating governments. As
to the agreement, monetary management problem must be solved by "diplomatic
negotiation," which greatly focuses on the point that, nobody is in charge of that
currency. The EMU does not work through good governance and unquestioned
political authority, rather through a vague spirit of shoddy compromise (The
Economist, 2011). Secondly, the territory of the Euro is likely to be confined to
the economies connected to the European Union while the process of circulation
of the dollar virtually is continued everywhere in the world.

The EU's share of the world market was as much as 25 percent in the 1990s
(calculated on a purchasing power parity basis) but it is gradually decreasing.
Three factors are working behind this considerable decline. First, the region's
demographics of the rapidly aging population which indicates a shortage of
labour force soon. Secondly, there is a reasonable prediction about the erosion of
human capital as the EU zone has been suffering a long unemployment problem
since 2008. Lastly, Europe's low rate of productivity growth plays an important
factor regarding this issue. Gros and Alcidi claim that the amalgamation of such
issues will deliver a weak GDP growth in Europe. The dimness of European
growth does not allow an adequate rate of job creation to ensure a rapid reduction in the unemployment rate. (Gros and Alcidi, 2013)

A major reason for the dollar's continued dominance is unlike the United
States, there is no single market for public debt in EMU. The eurozone's government debt market is cornered by the existence of differential credit and liquidity
risk premiums among member countries.

This brings us to our last point about Euro's unpromising appearance compared
to greenback. Europe cannot compete with America's unparalleled advantages

in issues of foreign policy ties or military reach. Of course, Europe has political stability and resource. But the military power which is required to attract outside investors cannot possibly be provided by EMU. Defence remains primarily the domain of individual sovereign states as "rapid reaction force," mainly exists on paper. So, each sovereign states highly acts according to their interest followed by EU's disability to create any sort of merged and effective armed units.

The renminbi

Several scholars and analysts argued that the renminbi (or yuan) be considered as the currency of the future and would be the true challenger of the dollar. Subramanian claimed that, China's increasing size of economic dominance is likely to interpret into currency dominance, the renminbi could exceed the dollar as the leading reserve currency well before the middle of the next decade (Subramanian, 2011).

However, the appeal of the yuan can be misleading. Recently the prospects of the yuan are no way more promising than the greenback for at least two reasons. First, China's mighty economic size has started to look fragile most recently. And then, the renminbi does not offer a large transactional network than dollars. The elaboration of these two points might be very important.

The enormous economy of Chinese has gained a leading position in export and has the second-largest market in imports. China is now the biggest trading partner of more than a hundred countries in the world. The middle kingdom has become a giant in global commerce beyond any doubt. But there is obvious doubt about the longevity of this trump card known as the giant economic size. Can China knockdown the "Second American Century" and start the "Chinese Century?" To find the answer, we need to examine the political economy of the renminbi.

Germain and Schwartz argue that when a domestic currency is wanted to attain the ability to becoming primary global currency, the issuer's stable domestic political institution plays a significant role (Germain and Schwartz 2014). It is because these institutions can create new demand for the currency in the world market through trade, current account, and balance of payments deficits, along with the capacity to ameliorate the costs that arise from those deficits. China has considerable institutional vulnerability regarding the mentioned issues. There is evidence that can show not only China's structural problems but also its dependency on growth in the US market. First, the World Bank calculates that final consumption of Chinese households fell from 46 percent of GDP in 2000 to 37 percent in 2014 (Zhang 2016). Every other Asian developing country is staying above China on that list. Although real household consumption nearly tripled from 1999 to 2013, and real GDP per capita nearly doubled from 1999 to 2008, but consumption growth fell short of supply growth. Then, the export has risen from 20% in 1998 to 39% in 2008 of China's GDP (Zhang et al. 2018). Net export contained one-third of the growth till 2008.

The two pieces of evidence proves China's repressed financial system and its political economy. Chinese financial system apprehends household savings and makes it flow towards state-owned or controlled firms and corporations. Individual savers had almost no option until recently but to either deposit money in banks which are state controlled with negative interest rates or to invest in the economy's over-relied sector of real estate. These state-owned firms capture various degrees of monopoly power so that they can earn profits. It is proved by the huge rise in Chinese savings from 37% of GDP in 2000 to 50% in 2007 which was produced by an increase in the state share of savings from 2.6% of GDP in 1999 to 21% in 2008 (Zhang 2016). The SOEs (State-owned firm) rarely contribute to this profit in mass domestic consumption. Rather these firms are channelling away the profits to create a real estate bubble. This kind of malinvestment is grossly overlooked by the government because of the connection between the SOEs and elites of the ruling communist party indicating the insider skim of approximately 1 trillion dollars. Thus, the ability of China's household to consume is reduced and so is the import.

Moreover, when the "Iron Rice Bowl" which is a system of employment-based welfare is connected to the regulation of hukou, it creates pressure on Chinese households to self-insure by saving large proportions of their income. The hukou system makes a million numbers of internal migrants in China to have lack access to the welfare state in the place where they currently work. All these factors lead to the tendency towards over-saving and under-consumption.

Thus, China has created a structurally biased economy, alarmingly leaning on the creation of over-capacity and excess export. As a result, China is facing difficulty earning the renminbi the status of international currency because China cannot flow the world with enough renminbi to effectively expand global demand for it. Of course, China could lend the other countries with renminbi, but this only creates the balance of payments surplus as that renminbi return to China for the purchase of Chinese products by the outside world. This is not an encouraging situation for foreign investors. For example, outsider firms have to be sure about the renminbi cash flow to accept the currency and this cash flow can only be ensured from Chinese demand for foreign goods and services. As with US trade surpluses in the 1920s, China put efforts to recycle that demand which has taken the form of lending to the United States via the purchase of Treasury and Agency debt. This lending merely reinforces the US dollar's pre-eminent position, instead of stepping toward the internationalization of the renminbi.

By 2015, there was a clear sign that the phenomenal growth of the Chinese economy is slowing down. China is still considered as a lower middle-income country even though it might provide soon the world's largest GDP on per capita basis. Comparing to the USA there is a presence of less sophistication in the Chinese economy. Their export products are mostly limited to low-quality consumer products with poor brand value. Political scientist Beckley argues that the United States has not declined; in fact, it is now wealthier, more innovative, and more militarily powerful compared to China than it was in 1991. China is rising, but it is not catching up. (Beckley 2012)

In the latest survey of global competitiveness by the World Economic Forum, China secured the 28th position, far behind the United States that is placing the third (World Economic Forum, 2014). Economists at the IMF suggest that China is now stepping towards "Lewis Turning Point" (Das and N'Diaye 2013). It provides a situation where a country will move from a vast supply of low-cost workers to sufficient labour unavailability.

But what about the important attributes for a currency to achieve primacy in the world market? In reality, China offers little appeal in finance for its conservative market policy for the outside world. A few international investors were granted direct access in the domestic capital market in 2002 but under the so-called Qualified Foreign Institutional Investor (QFII) scheme. A new initiative in 2002 was launched permitting indirect purchases of shares on the Shanghai stock exchange through the Hong Kong market. However, this new opportunity was under strict administrative control. Though the yuan has been added to the list of currencies that measures the value of IMF's Special Drawing Rights, it has a long way to go to be freely usable currency worldwide.

If we consider the factor of military reach, China's vast military buildup, which is designed to project power beyond the country's internal and external security. Instead of having this powerful military capability, China couldn't show less concern about the global security issue. Rather the country is now considered as aggressively working only for its core national interest. Moreover, among all previous cases of currency internationalization in the modern era, none has an authoritarian domestic political history like China. Even Beijing's absolute success in terms of macroeconomic management cannot produce confidence in investors' mind. It is because of their little concern for property rights or faithful enforcement of contractual obligations.

The dictatorial nature of the ruling political party creates doubt about their respect for the rule of law and maintaining the standard of effective governance. China ranks 82nd for the rule of law in the survey about global governance by The World Bank (WJP 2019). More shockingly, according to Transparency International, China secured the 80th position among all the 198 countries in the corruption index (Transparency International 2020). All of these lead us to the critical commentary Lo that China faces a credibility problem (Lo 2013).

Prospect of China's digital currency

China's push to introduce a digital currency began in 2014. After 5 years, in December 2019, the PBoC (People's Bank of China) became successful in introducing such a currency. Named the Digital Currency Electronic Payment (DCEP), the pilot work included the Big Four state banks in Shenzhen and Suzhou, as well as the Big Three telecoms firms. Beijing has now added more cities to the experiment (Xie 2020).

Experts have argued that China's desire to have a crypto currency can be seen as part of its larger effort to assert or boost global reserve currency pool dominance and advance foreign policy claims. A digital renminbi could make a dent into the United States' ability to pursue national interests by leveraging the US dollar. A Chinese crypto renminbi could rapidly challenge the dominance of the US dollar.

In the broader context, the creation of DCEP is a part of the "Chinese Dream" of President Xi Jinping to build the global influence of China. A digital renminbi could advance the currency's internationalization process and boost the Belt and Road Initiative simultaneously.

Currently when two companies that use different currencies sign a business contract, the value of the transactions is usually denominated in the US dollar. This means they need a bank to act as an intermediary. But with digital currency, this will not be the case anymore.

A foreign importer and a Chinese exporter could instruct computers in a blockchain network to place a renminbi-denominated payment as a new way of doing business. This decentralized escrow system could allow foreign companies to negotiate trade deals without using dollars to hedge their exchange rates risk. And this digital currency can be used all around the world even for everyday transactions.

Experts have also noted that DCEP bypasses the Western banking system, including SWIFT.

In time, Beijing could offer DCEP-based direct machine-to-machine payments along its 68-country Belt and Road Initiative which covers 65% of the world's population and 40% of the global gross domestic product (Duarte and Leandro 2020). It could also encourage its allied African governments to digitally peg the value of their domestic currencies to the DCEP.

However, not all thinkers share the same view about the prospect of the new Chinese digital currency. Henry M. Paulson Jr., former US Secretary of the treasury, is one of those experts who thinks that the effects of China's digital currency will be minimal and that it will not cause any more damage to the US dollar centric financial structure than the RMB itself. Despite the renminbi going digital, the basic mechanisms of reserve currencies will still be dictated by the relative openness of the political systems and the demand for a particular currency as a virtue of the function of its dependent economy.

Conclusion

The dollar has a long-standing future and primacy over all the currencies which challenged it. The five prerequisites for successful currency internationalization provide the explanation of it. Only the dollar embodies the full list of attributes which makes it the fittest currency to have control over the world market. To be an effective rival of the greenback, a currency must have better credentials than the Euro or the yuan. That means the Euro or Europe's money will need better governance structures and reintegration of EMU's balkanized financial sector.

China's RMB, however, will need more financial development. Beijing has to be more sophisticated in terms of foreign policy and domestic political economy model. Considering the current geopolitical situation, it is not an easy task to bring the recommended changes for both EMU and Beijing. Therefore, the survival of the fittest is the most reasonable forecast here. The world economy might have to go a very long way to witness the true fall of the dollar.

References

Aliber, R. Z. (2002), *The new international money game*, 6th edition, Chicago: University of Chicago Press.

Bacchetta, P and Van Wincoop, E (2005), A theory of the currency denomination of international trade, *Journal of International Economics*, December.

Bayoumi, T. and Eichengreen, B. (1998), *Exchange rate volatility and intervention: implications of the theory of optimum currency areas*. https://doi.org/10.1016/S0022-1996(98)00032-4

Beckley, M. (2012), China's century? Why America's Edge will endure. *International Security, 36*(3), pp. 41–78.

Calleo, D. (2009), Twenty-first century geopolitics and the erosion of the dollar order, in Eric Helleiner and Jonathan Kirshner (eds.), *The Future of the Dollar*, Ithaca NY: Cornell University Press, 164–190.

Chinn, Menzie and Frankel, Jeffrey (2008), The euro May over the next 15 years surpass the dollar as the leading International currency, *International Finance* (forthcoming).

Cohen, B. J. (2015), *Currency Power: Understanding Monetary Rivalry*, Princeton NJ, Princeton University Press.

Cohen, B. (2009), Toward a Leaderless Currency System. In Eric Helleiner and Jonathan Kirshner, eds., *The Future of the Dollar* (142–63). Ithaca, NY: Cornell University Press.

Chey, H. (2012), Theories of International currencies and the future of the world monetary order. *International Studies Review, 14*(1), 51–77. Retrieved from www.jstor.org/stable/41428882

Das, M. and N'Diaye, P. (2013), *Chronicle of a Decline Foretold: Has China Reached the Lewis Turning Point?*, Working Paper WP/13/26, Washington, DC, International Monetary Fund.

Dawnay, A. (2001). A history of sterling. Retrieved from https://www.telegraph.co.uk/news/1399693/A-history-of-sterling.html

Duarte, A. P. B. and Leandro, J. F. B. S. (2020). *The Belt and Road Initiative: An Old Archetype of a New Development Model* (1st ed. 2020 ed.) Palgrave Macmillan.

Duffin, E. (2020), *Top 20 export countries worldwide in 2019*. Retrieved from https://www.statista.com/statistics/264623/leading-export-countries-worldwide

Flandreau. (2008), The rise and fall of the dollar, or when did the dollar replace sterling as the leading international currency? National Bureau of Economic Research. Retrieved from http://www.nber.org/papers/w14154

Eichengreen, B. (2011), *Exorbitant Privilege: The Rise and Fall of the Dollar and the Future of the International Monetary System*. New York: Oxford University Press.

Eichengreen, B and Flandreau, M. (2008), *The Rise and Fall of the Dollar, or When Did the Dollar Replace Sterling as the Leading International Currency?* Retrieved from https://www.nber.org/papers/w14154

Engel, C. (2005), *Equivalence results for optimal pass-through, optimal indexing to exchange rates, and optimal choice of currency for export pricing, NBER Working Paper*, no 11209.

Fields, D. and Vernengo, M. (2013), Hegemonic currencies during the crisis: The dollar versus the euro in a cartelist perspective, *Review of International Political Economy* 20: 4(August),740–759. DOI: 10.1080/09692290.2012.698997

Frankel, J. A. (1995), Still the lingua Franca. *Foreign Affairs*, 74, pp. 4 (July/August), 9–16.

Germain, R. and Schwartz, H. (2014). The political economy of failure: The euro as an international currency. *Review of International Political Economy, 21*(5), pp. 1095–1122.

Grey, C. S. (2019). The United States and the World Order. Retrieved from https://www.nipp.org/2019/02/06/gray-colin-s-the-united-states-and-world-order/

Gros, D. and Alcidi, C. (2013). The Global Economy in 2030: Trends and Strategies for Europe. Retrieved from https://espas.secure.europarl.europa.eu/orbis/sites/default/files/generated/document/en/The%20Global%20Economy%20in%202030.pdf

IMF. (2020). Currency Composition of Official Foreign Exchange Reserves (COFER). Retrieved from https://data.imf.org/?sk=E6A5F467-C14B-4AA8-9F6D-5A09EC4E62A4

Kempe, F. (2019), A geopolitical earthquake has shaken US leadership in the world — Russia and China stand to benefit. *Retrieved from* https://www.cnbc.com/2019/10/26/russia-and-china-are-challenging-us-global-leadership-under-trump.html

Kenen, P. B. (2011). Currency internationalisation: an overview. In: BIS Papers chapters. RePEc:bis:bisbpc:61–04. Retrieved from https://ideas.repec.org/h/bis/bisbpc/61-04.html

Kindleberger, C. P. (1976), Lessons of floating exchange rates," Carnegie-Rochester Conference Series on Public Policy, Elsevier, vol. 3(1), pages 51-77, January.

Lee, J. (2019). Why the Almighty US Dollar Will Remain the World's Currency of Choice. Retrieved from https://www.hudson.org/research/15128-why-the-almighty-us-dollar-will-remain-the-world-s-currency-of-choice

Li, Q. (2003). The effect of security alliances on Exchange-rate regime choices. *International Interactions* 29:2, 159–193. DOI: 10.1080/03050620304600

Lim, E. (2006). The Euro's Challenge to the Dollar: Different Views from Economists and Evidence from COFER (Currency Composition of Foreign Exchange Reserves) and Other Data.

Lindert, Peter H. (1969), *"Key Currencies and Gold, 1900–1913," Princeton Studies in International Finance No. 24*, International Finance Section, Department of Economics, Princeton University.

Lo, C. (2013), *The Renminbi Rises: Myths, Hypes and Realities of RMB Internationalisation and Reforms in the Post-Crisis World*, London: Palgrave Macmillan.

Mundell, R. A. (2000), "The euro and the stability of the International monetary system," in R. A. Mundell and Cleese, A. (eds.), *The Euro as a Stabilizer in the International Economic System* (Boston: Kluwer Academic), 57–84. DOI: 10.1007/978-1-4615-4457-9

Oatley, T., Kindred Winecoff, W., Pennock, A. and Bauerle Danzman, S. (2013). "The political economy of global finance: A network model. *Perspectives on Politics*, 11, pp. 133–153.

Posen, A. (2008), Why the euro will not rival the dollar, *International Finance* 11:1 (Spring), 75–100. Retrieved from https://onlinelibrary.wiley.com/doi/full/10.1111/j.1468-2362.2008.00217.x

Rajendran, G. (2013), Financial blockades: Reserve currencies as instruments of coercion, in Wheatley, A. (ed.), *The Power of Currencies and Currencies of Power*, London, International Institute for Strategic Studies, 87–100.

Sobel, A. C. (2012), *Birth of hegemony: Crisis, Financial Revolution, and Emerging Global Networks*, Chicago, University of Chicago Press.

Starrs, S. (2013), American economic power Hasn't declined – It globalize! Summoning the data and taking globalization seriously. *International Studies Quarterly*, 57(4), pp. 817–830.

Subramanian, A. (2011), *Eclipse: Living in the Shadow of China's Economic Dominance.* Washington: Peterson Institute for International Economics.

Tett, G. (2014), "Why the dollar stays steady as America declines?" *Financial Times.*

The Economist. (2013). The euro crisis: The sleepwalkers https://www.economist.com/leaders/2013/05/25/the-sleepwalkers

The Economist. (2011). After the crisis: Making do. Retrieved from https://www.economist.com/taxonomy/term/76972/www.marjoriedeane.com?page=2152

Triffin, Robert (1960), *Gold and the Dollar Crisis*, New Haven: Yale University Press.

Usdebtclock. (2020). US National Debt. Retrieved from https://www.usdebtclock.org/

Vine, D. (2015). Where in the World Is the U.S. Military? Retrieved from https://www.politico.com/magazine/story/2015/06/us-military-bases-around-the-world-119321#:~:text=Despite%20recently%20closing%20hundreds%20of,about%2030%20foreign%20bases%20combined.

Wheatley, A. (2013). Introduction, in Wheatley, A. (ed.), *The Power of Currencies and Currencies of Power*, London, International Institute for Strategic Studies, 9–16. DOI: 10.1080/01440350208559416

Xie, J. (2020). China's Digital Currency Takes Shape; Will It Challenge Dollar? Voice of America. Retrieved from https://www.voanews.com/economy-business/chinas-digital-currency-takes-shape-will-it-challenge-dollar

Zhang, L., Brooks, R., Ding, D., Ding, H., Lu, H. H. J. and Mano, R. (2018). China's High Savings: Drivers, Prospects, and Policies.

Zhang, L. (2016). Rebalancing in China – Progress and Prospects. Retrieved from https://www.imf.org/external/pubs/ft/wp/2016/wp16183.pdf

Transparency International. (2020). Corruption Perception Index – 2019. Retrieved from https://www.transparency.org/en/cpi/2019/results/chn

Richter, W. (2018). Here's why efforts to reduce the dollar's dominance won't work anytime soon. Retrieved from https://www.businessinsider.com/us-dollar-heres-why-it-will-remain-dominant-around-the-world-2018-7

Tavlas, G. S., and Ozeki, Y. (1992), The Internationalization of Currencies: An Appraisal of the Japanese Yen. Occasional Paper 90. Washington: International Monetary Fund.

The balance. (2020). Will the US Dollar Collapse? How, When, and What to Do If That Did Occur. Retrieved from https://www.thebalance.com/when-will-the-u-s-dollar-collapse-3305691#:~:text=Effects%20of%20a%20Dollar%20Collapse,prices%20would%20skyrocket%2C%20causing%20inflation.).

The Northern Miner. (2013). Commentary: Ned Goodman and the "Botox Economy," pt. 2. Retrieved from. https://www.northernminer.com/commodities-markets/commentary-ned-goodman-and-the-botox-economy-pt-2/1002620584/

WJP. (2019). Rule of Law Index. Retrieved from https://worldjusticeproject.org/sites/default/files/documents/ROLI-2019-Reduced.pdf

World Bank. (2011), Multipolarity: The New Global Economy. Washington: World Bank.

World Economic Forum. (2014), The Global Competitiveness Report 2014–2015. Retrieved from http://www3.weforum.org/docs/WEF_GlobalCompetitivenessReport_2014-15.pdf

8

DREAM AND REALITY OF FREE MOBILITY IN ASIAN LABOR MIGRATION REGIMES

Kazue Takamura

Introduction

This chapter aims to examine the *intricate interactions* between global markets, labor mobility, and the costs of migration. While the increased transnational labor mobility, especially in low-skilled forms, is seen as an integral seed of socio-economic freedoms for the poor, scholars and human rights advocacy groups have also documented myriad risks and costs that are structured in the contemporary labor migration regimes (Xiang and Lindquist 2014; Lan 2006; Rodriguez 2010; Geiger 2013). The assumed positive externalities of international labor migration have been increasingly debated in the literature of migration and development. The costs of migration are evident in terms of long-term family separation, health hazards associated to migrants' work environments, human trafficking, exploitative employers, and highly abusive agencies that charge excessive pre-departure fees (The World Bank 2016, p. 15). By engaging with the critical literature of migrant mobility and control, I will examine the ways in which Asian labor migration regimes are characterized by their highly contradictory policies. I argue that Asian migration regimes embrace both *liberal* labor migration that allows more unskilled workers' transnational labor mobility and *illiberal* migrant surveillance that ensures migrant workers' disposability, exploitability, and unprotection. Asian intraregional labor migration merits a serious analytical attention given its profound heterogeneity. This is apparent in its roles (labor-sending, labor-receiving, and both), labor-import policies (official, side-door, and back-door), migratory flows (legal temporary foreign worker programs, irregular channels, non-worker student, and marriage visas), labor market demands (agricultural, seafood, industrial, construction, education, health, care, and service sectors), and skills (professionals, medium-skills, and low-skills) (Baas 2016; Liu-Farrer and Yeoh 2018; Asis and Piper 2008). A critical lens of Asian intraregional migration unpacks the growing imbalances between the free movement of capital and the barriers against labor mobility in the age of the global free market.

The chapter will be divided into three sections. In the first section, I will interrogate the promotion of labor migration as triple wins in the literature of international development. In the second section, the chapter will examine the intricate interconnection between neoliberal free markets and migrant surveillance. I will also highlight the particularities of Asian intraregional labor migration. In the third section, I will examine the effects of neoliberalism in terms of normalizing migrants' perpetual immobility. In particular, I will engage with the recent critical studies of Asian labor migration and mobility surveillance. I will provide an empirical analysis of Asian labor migration regimes based on Japan's labor import policies (foreign technical interns) as well as on the Philippine's labor-export (labor brokerage) system. This chapter will exhibit the ways in which labor-receivers, labor-senders, and non-state actors collectively contribute to the normalization of migrant immobility while promoting increased and flexible labor mobility.

Labor migration as triple wins

According to the International Labor Organization's estimation published in 2018, there were a total of 163.8 million migrant workers globally. Asia is one of the busiest migratory corridors in terms of both supplying and absorbing migrant workers. Asia and Pacific received 20.4% of the global migrant worker population (International Labor Organization 2018). This number does not include those who are irregular, meaning officially not being registered by host governments. The inflow of remittances is relatively stable and expanding compared to the fractured flow of foreign direct investment. The total amount of the global remittance money that flows into low and middle income countries is three times higher than the amount of development aid (The World Bank 2019, p. vxii). According to the World Bank, the low and middle income countries in "East Asia and Pacific" received a total of 149 billion US dollars in 2019 (ibid., p. 15). This number counted for **one fourth** of the global remittances. The total inflows of remittances accounted for 6.5% of Vietnam's GDP (The World Bank 2020-a) and 9% of the Philippines' GDP in 2019 (The World Bank 2020-b). The expansion of *intraregional* labor mobility is particularly evident for ASEAN (Association of Southeast Asian Nations) countries. For example, three among the top five destination countries for the ASEAN migrants were Thailand (3.6 million), Malaysia (1.5 million), and Singapore (1.2 million) in 2013 (International Labor Organization 2015-a). The World Bank report in 2019 predicts a continuous and significant increase of global migratory flows due to the widening income gap between high-income and low-income countries (The World Bank 2019).

The promotion of temporary and circular labor migration is oftentimes seen as a ***triple-win*** policy for labor-sending countries, labor-receiving countries, and

migrants themselves (Curtain et al. 2016). Major international financial institutions and policy makers have promoted labor migration by quantifying its positive
externalities of remittances on development. The export of labor is considered as
a prerequisite for sustainable development in the Global South. According to the
World Bank, "economic development and growth require people to move where
the jobs are…. migration offers the best opportunity for finding a better job and
thereby escaping poverty and unemployment. (ibid., p. 1). It is generally suggested
that key causes of labor migration are rooted in poverty and wage gaps across
countries (ibid., p. 6). Low-income countries gain from remittances in terms
of paying off accumulated foreign debt and eradicating chronic poverty. Labor
migration has been viewed as an impactful development tool for bringing positive
externalities on economic, social, and political domains. It curbs high unemployment rates, improves living standards of the poor, and thus achieves overall
socio-economic stability. Migrants and their families experience positive impacts
of remittances in terms of increased income, improved food consumption, higher
school enrolment rates, reduced child mortality, and overall improved welfare at
home (The World Bank 2016, p. 15). Scholars point out that remittances serve
as primary insurance for left-behind families in the times of natural disasters and
other climate-related crises (ibid., p. 16). Remittances are thus considered as a
real "bottom-up way of redistributing and enhancing welfare among populations
in developing countries." (De Haas 2005, p. 1277). Remittances also generate
positive externalities for local businesses. Migration-related industries, including
recruitment agencies, remittance firms, travel agencies, and labor training and
education, have expanded sharply in low and middle-income countries in the
past two decades (Rodriguez 2010). Labor migration thus becomes an emblem of
"successful" development for the Global South.

For high-income countries, on the other hand, a labor-import policy is
considered as an affordable and convenient solution to overcome chronic labor
shortages in the low-wage labor-intensive sectors (Ronald 2012, p. 38). High-
income Asian countries, including Japan, Singapore, Taiwan, South Korea, and
Hong Kong, have increasingly shifted to the dependence on inexpensive foreign workers. Temporary migrant workers are expected to rescue the shrinking
small-and-medium sectors, including construction, manufacturing, food production, and care labor services. Advanced East Asian economies commonly
experience growing demands for affordable elderly care and health workers due
to the expanding aging population and the shrinking working-age population
(Song 2015; Lan 2018). Foreign workers thus serve as integral backbones of Asian
Tigers' sustained economic growth.

Argument – Neoliberal migrant surveillance

This chapter, however, will not seek to explain the expansion of Asian intraregional labor migration based on the logic of triple wins in development. It
will focus on the paradoxical interconnection between migrant mobility and

immobility. In order to unpack the mobility-immobility nexus that is embedded in the Asian intraregional labor migration regimes, I will pay attention the tools of *migrant surveillance*, that is operationalized by (a) labor-senders, (b) labor-receivers, and (c) the market (employers and intermediaries). I argue that the growing convergence between migrants' mobility and their intensified immobility cannot be fully explained by the conventional logics of push and pull factors or triple wins. It requires a careful attention to the mechanism of power and control that is embraced by the contemporary Asian labor migration regimes. I argue that migrant surveillance is a neoliberal technology of power that seeks to control the processes of deployment, employment, labor discipline, and repatriation of unskilled migrant workers (Xiang 2013; Lan 2006; Rodriguez 2010). The neoliberalization of Asian economies has led to the expansion of labor mobility through bilateral agreements and enlarged transnational flows of capital to the Global South. However, it also has normalized the commodification of migrant labor because states seek opportunities of cooperation to achieve their own neoliberal economic goals. State-level migration management tools are increasingly popular in the forms of bilateral agreements and Memorandum of Understandings (MoUs) (International Labor Organization 2015-b). Despite the official promotion of intraregional labor mobility, these bilateral documents "rarely provide for migrant worker protection" (Kneebone 2010, p. 384). For example, Indonesia (as a labor-sender) and Malaysia (as a labor-receiver) have signed bilateral MoUs to promote increased legal channels for temporary labor migration. These bilateral MoUs intend to curve the massive number of undocumented Indonesian workers in Malaysia (Arifianto 2009). However, as Kneebone claims, such bilateral agreements tend to prioritize "the interests of employers and do not prevent rights abuses" (Kneebone 2010, p. 387). Furthermore, scholars contend that bilateral agreements serve as a regulatory tool to achieve transnational policing and surveillance of migrants (Xiang 2013, p. 85). Thus, bilateral forms of promoting labor mobility unfortunately contribute to the commodification of, as well as the increased control of, migrant labor rather than the protection of workers. In short, *illiberal forms of* migrant surveillance are constitutive of the Asian labor migration regimes for their *liberal* market-centric economic goals.

Particularities of Asian intraregional labor migration

The mainstream migration literature largely focuses on the transnational migratory flows from the developing South to the developed North. Such a South-to-North migration lens overshadows the long-standing and dynamic labor mobility and transactions within the South (De Haas 2005, p. 1270). The scope of Asian intraregional labor mobility merits a serious analytical attention expanding transactions of capital and labor within the region. There are four particularities illustrative to the Asian intraregional labor migration regimes. First, the Asian region stands out because of its economic diversity. The rapid industrialization of the East Asian and Southeast Asian economies since the late 1970s has led to the expansion of intra-regional migration. The availability of inexpensive and disposable workers in neighboring countries is a crucial determinant for the advanced Asian economies to maintain their competitive productivity. Intraregional labor supply also provides a solution to the chronic labor shortages in the low-wage manufacturing and agricultural sectors. Second, Asia's top destination countries are middle-income countries. Furthermore, these countries are also labor-exporters. For example, Malaysia and Thailand are listed among the top five destinations in the region (International Labor Organization 2015-a). These countries have also served as labor-senders to the Gulf States, as well as to the high-income Asian economies, since the mid-1970s.

Third, increased labor dependency does not always imply the liberalization of immigration that is assumed in the Global North. For example, Japan is one of the major advanced Asian economies that is excessively depending on low-skilled migrant workers from neighboring Asian neighbors while maintaining restrictive immigration policies that prohibit these migrant workers to become long-term residents. The government has implemented various "side-door" policies to import cheap and disposable workers in the forms of "trainees/interns," international students, and co-ethnic Nikkei migrants (Yamanaka 1993; Yamamoto and Johnson 2014; Akashi 2017). Fourth, Asia embraces highly divergent political regimes, ranging from liberal democracy to illiberal military and authoritarianism (Liu-Farrer and Yeoh 2018, p. 1). This political diversity challenges the migration model that assumes the linkages between greater labor mobility, economic growth, and liberal democracy. For example, the Ferdinand Marcos dictatorship in the Philippines had institutionalized the system of mass labor export, or a "labor brokerage system," in the mid-1970s (Rodriguez 2010). As I will discuss later, the Philippines' expansion of labor export was rooted in the United States' Cold War strategy to eliminate the seeds of communism through development aid and the promotion of remittances (ibid.). Indonesia also took the similar labor export path as the Philippines during the Suharto's authoritarian regime in the 1980s (Silvey 2004). The distinct convergence between illiberalism, the Cold War geo-politics, and the increased labor mobility thus cannot be fully explained by a singular labor migration model that is largely based on the South-to-North mobility.

Migrants' perpetual immobility

In this section, I will discuss the critical linkages between the neoliberalization of Asian economies and migrants' perpetual immobility. **Mobility** signifies the increased labor mobility therefore better income opportunities for low-skilled migrant workers. Labor mobility generates other positive externalities on human development including health, education, and livelihoods. **Immobility**, on the other hand, implies unfreedom and risks. Low-skilled temporary migrant workers experience myriad forms of risks in terms of financial insecurity, labor exploitation, devaluation of labor, and the arbitrary termination of their labor contract.

Furthermore, as NGOs and the media have documented, migrant workers tend to bear accumulated financial burdens due to the highly abusive pre-departure broker fees as well as remittance dependence at home. Migrant workers are thus squeezed between the normalization of labor rights violations and the perpetual economic burdens at home. Furthermore, the contemporary patterns of labor immobility are highly gendered. The expansion of Asian labor migration regimes is deeply rooted in the feminization of labor. Scholars have investigated the distinct gendered patterns of migration, including domestic workers, elderly caregivers, entertainers, health professionals, or marriage migrants (Lan 2006; Silvey 2004; Constable 2009; Silvey and Parreñas 2019). Such gendered labor mobility crucially structure these workers' vulnerability. Silvey and Parreñas' work on the migrant domestic workers to the Gulf States unpacks gendered forms of immobility. Migrant women experience "chains of precarity" that are exacerbated by "their indebtness and dependency on a recruitment agency to determine both their employer and their country of destination (the precarity of migration)," limited labor protection (the precarity of labor), and the protracted income insecurity (the precarity of future) (Silvey and Parreñas 2019, p. 2).

I argue that such unspeakable and perpetual pain of migrant immobility is further sealed by migrant surveillance operationalized by the contemporary Asian migration regimes. Migrant surveillance is an essential part of the neoliberal labor mobility system that ensures labor productivity, labor discipline, and labor disposability. While a neoliberal market economy promotes *reduced* regulation of the market by the state, it ironically *strengthens* the state's power to police and regulate transnational labor mobility (Kretsedemas et al. 2017, p. 4). The recent expansion of immigration detention and deportation explains the profound immobility of migrant workers in the age of neoliberal labor mobility (ibid.). Scholars argue that migrant surveillance and punishment allow both the state and the market to manufacture an army of productive, cheap, and disciplined workers who support the success of global market economies. For example, Peichia Lan argues that female migrant workers' mobility is constrained due to the *"bonded* global market" that ensures temporary, regulated, and disciplined forms of labor mobility between labor-sending and labor-receiving countries. Lan argues that the formation of a "bonded labor market" is integral for Asian advanced economies to make domestic production globally competitive and financially affordable (Lan 2006, p. 35).

The neoliberal chains of migrant immobility could be further explained by the golden rule of "compulsory return" (Xiang 2013). According to Biao Xiang, compulsory return serves as a technology of migrant control that is built into the contemporary Asian temporary foreign worker programs (ibid., p. 85). Xiang elaborates on the distinct mechanism in which compulsory return is institutionalized by a "spectrum of actors located at different levels, transnational, national, and local" (ibid.). A "transnational space of surveillance and policing" is thus established and managed not just by states (based on bilateral agreements) but also non-state actors, including intermediaries and employers. The primary goals of such transnational surveillance are to isolate migrants from their host communities as well as to deprive migrants from claiming their rights and benefits" (ibid., pp. 84 – 85). The transnational space of surveillance attests that the expansion of labor mobility is, in reality, mediated and conditioned and thus subject to regulation. It does not lead to migrants' increased freedom and capability. Instead, it transforms "migration into an object of intensive regulation, commodification, and intervention." (Xiang and Lindquist 2014, p. 127). The construction of structured migrant immobility thus requires a set of technologies as well as actors who separately or collectively impose control and surveillance on migrant bodies. The expansion of the contemporary Asian intraregional labor migration regimes is deeply interconnected to the highly abusive technology of migrant surveillance that maintain migrant workers' labor disposability and exploitability. Thus, it sustains perpetual chains of migrant precarity (Silvey and Parreñas 2019).

Japan – Neoliberal commodification of migrant labor and surveillance

A "transnational space of migrant surveillance" (Xiang 2013) is evident in Japan's "foreign technical interns" whose labor disposability is strictly monitored by the host country, the home country, the employer, and the recruitment agency in the home country. Since the late 1980s, Japan has intensified its excessive dependence on migrant workers especially in the small and medium firms. The number of officially employed foreigners increased nearly a twofold from 686,000 in 2011 to 1.66 million in 2019 (The Ministry of Health, Labor and Welfare, Japan 2019). These non-citizen workers are not just permanent residents but also temporary residents including trainees, international students, and professionals (ibid.). Despite this visible dependency on foreign workers, the Japanese government is reluctant to introduce official foreign worker programs. Instead, the government has implemented a set of "side-door" immigration programs to temporally respond to the severe labor shortages due to the rapidly greying population (Chung 2010, Yamanaka 1993). As Junichi Akashi claims, one of the most distinct characteristics of contemporary Japan's immigration policies is the convergence of the expanded absorption of foreign labor and the apparent rejection of immigration (Akashi 2017, p. 12). The Technical Interns Training Program (TITP) serves as Japan's integral *de facto* guest worker program. The original model was implemented back in 1990 (Yamanaka 1993). The recent expansion of this "side-door" program is especially notable. Under the TITP, Japan received more than 380,000 intra-regional temporary foreign workers in 2019 (The Ministry of Health, Labor and Welfare, Japan 2019). This number was four times higher compared to the year 2009 (The Ministry of Justice, Japan 2010). The main labor source countries were Vietnam, China, Indonesia, the Philippines, and Thailand in 2019. Scholars argue that Japan has obtained a large pool of low-cost and disposable workers through side-doors and back-doors without liberalizing immigration policies (Akashi 2017; Bélanger et al. 2011). The TITP is presumably grounded on Japan's paternalistic development ideology – a false development promise based on an educational category such as a "trainee" who is supposed to gain skills and techniques– with the inclusion of rights and protection *at least* on the paper. It however produces a large segment of cheap and dependable foreign workers who are not considered, not fully protected, as "laborers." The TITP is heavily criticized as being "synonymous with a gamut of human rights violations" (The Japan Times 2016). A human rights lawyer, Shoichi Ibusuki, contended that the TITP revisions in 2016 did not address the fact that trainees are continuously denied the discretion to change their workplaces no matter how much they are abused or underpaid (ibid.). Foreign technical interns are not just susceptible to labor exploitations but also highly vulnerable to the loss of status due to the involuntary labor contract violations, such as work-related injuries and illnesses. The TITP's employer-tied visa system normalizes migrants' risk of losing their jobs thus their legal status. Such legal arrangement expands the employer's excessive control over migrant workers not just limited to workers' labor discipline but also workers' everyday lives beyond work. The excessive power of labor discipline invites myriad of labor rights abuses. These rights violations include excessive overtime work hours, unpaid or underpaid wages, coercive salary withholding under the name of "mandatory savings," passport confiscation, and habitual forms of verbal and physical violence by their employers. Oftentimes, employers impose coercive power to control foreign interns' freedom by imposing rules and penalties (Gaikokujin Kensusei Network 2006; Bélanger et al. 2011). Because of the profound fear of loss of status, migrant workers endure abusive labor treatments and unprotection rather than demanding for fair labor treatments. Even those who seek to run away from abusive employers would be instantly subject to criminalization (illegality) and punishment (deportation) due to the employer-tied visa policy (Gaikokujin Kensusei Network 2006). The number of runaway foreign interns is rising with the expansion of the program. The Immigration Services Agency, the main government body of immigration control, problematizes the recent increase of "missing" foreign interns who are now under the category of "illegal aliens." (Immigration Services Agency, Japan 2019). The increase of runaway interns is seen as a direct threat to the golden rule of compulsory return that is central to Japan's neoliberal pursuit of market profitability (The Japan Times 2019). In order to deter migrants from running away before completing their labor contract, both states and non-state actors collectively form a transnational space of policing and surveillance (Xiang 2013). For example, a commercial recruitment agency in Vietnam would impose an expensive "deposit" to a prospective migrant worker in order to prevent the worker from breaking one's labor contract and from refusing to return home after the contract ends in Japan (The Mainichi Shimbum 2019). Migrant surveillance tools, implemented by various state and non-state actors, are to ensure that the migrant worker will return home once the contract ends and the worker will not claim one's rights and entitlements that are exclusive to citizens and long-term residents. Detention and deportation by the labor-receiving state as well as the excessive financial burdens imposed by middlemen are the integral surveillance tools to discipline migrant workers who seek to challenge the fundamental neoliberal rationale of labor disposability and exploitability (Kretsedemas et al. 2017, p. 3).

The Philippines – A labor brokerage state

While scholars focus on the securitization of border and migrant surveillance imposed by labor-receiving countries, one should also pay attention to the role played by labor-sending countries in further contributing to the formation of a transnational space of migrant surveillance (Xiang 2013; Rodriguez 2010). Although a labor-export policy is part of the neoliberal economic package that aims to reduce the role of the state in development, it ironically encourages the state's authoritative role in facilitating and regulating the transnational out-flows of labor. This is because the regulation of labor recruitment, deployment, and most crucially, migrants' return are all integral to ensure the sustained inflows of remittances as well as sustained opportunities for temporary labor migration (Xiang 2013). Legal and temporary forms of labor mobility are thus not just agendas for labor-receiving countries to ensure their regulated labor-import policies, but also are important priorities for labor-sending countries to achieve their own economic goals. Robin Rodriguez' compelling work on the Philippines as a labor brokerage state suggests the critical regulatory role of the labor-sending state in terms of monopolizing the authority to mobilize and reg-ulate the transnational movements of citizen workers (Rodriguez 2010, p. xxiv). The Philippine state has implemented a highly bureaucratic labor export policy since the mid-1970s under the dictatorship of Ferdinand Marcos who introduced neoliberal economic policies. The country's implementation of the structural adjustment programs (SAPs) was considered as a solution to its accumulated debt as well as to the mounting unemployment rates (ibid., p. 12). The introduction of SAPs was also deeply political as the Philippine state received development aid and IMF loans as part of the United States' anti-communist Cold War strategy in Southeast Asia (ibid.). The Philippines' labor brokerage system (ibid., p. x) was born out of these urgent economic and political needs to implement neoliberal economic policies. According to Rodriguez, labor brokerage is "a neoliberal strategy that is comprised of institutional and discursive practices through which the Philippine state mobilizes its citizens and sends them abroad to work for employers." (ibid.). The labor brokerage system thus maximizes financial reve-nues based on citizens' economic commitment to send money back home. This system is integral to facilitate the stable outflows of labor as well as to ensure the sustained inflows of remittances. The success of labor brokerage requires not just institutional efforts to facilitate labor opportunities abroad for citizens (for example, through bilateral agreements), but also the discursive promotion of productive and disciplined "ever-willing workers" (for example, through a nationalistic rhetoric of "national heroes") (ibid.). In many low-income or lower middle-income countries, remittances increasingly become a primary resource of neoliberal welfare for those who do not have access to affordable public welfare. Neoliberal labor migration promotes the poor to be self-responsible and "empowered" by sending and receiving remittances. Labor migration is thus considered as effective and self-reliant development machine for the poor in the age of neoliberalism.

Conclusion

This chapter attempted to provide a critical analysis of Asian labor migration regimes by focusing on the contradictory conjunction of labor mobility and immobility. The neoliberalization of Asian economies intensifies the triple forms of structural dependencies. These are (1) the dependency of labor-sending countries on remittances as well as on other migration-related business revenues, (2) the dependency of labor-receiving countries on intraregional temporary migrant workers and on their low-cost but higher productivity, and (3) the dependency of migrant workers on temporary, contract-based, and low-wage labor mobility. I argue that these triple dependencies are central to explain the paradoxical interaction between labor mobility and structured immobility in Asia's intraregional labor migration. These dependencies are deliberate consequences of the neoliberalization of Asian economies. The self-generating and protracted patterns of the triple dependencies further solidify the paradoxical mobility-immobility nexus. In short, the "wellbeing" of Asian high-income economies is promised by an abundant supply of cheap and disposable migrant workers from Asian low-income countries. While neoliberal market economies celebrate the expanding income opportunities for the poor, these same policies also ensure to maintain migrant workers' vulnerability. In short, migrant immobility is not an unintended or accidental but a *deliberate and structured* outcome of the formation of Asian intraregional labor migration regimes. The study of a migrant mobility-immobility nexus provides a critical lens to understand the profound contestations and contradictions in the global expansion of "free" markets. This chapter interrogated the manner in which the increased "freedom" of the global market economy goes hand in hand with the illiberal practices of coercive labor discipline, exploitation, and punishment that sustain "unfreedom" of migrants.

References

Akashi, Junichi (2017) *Abeseiken no gaikokujin seisaku* (abe administration's immigration control policies) *Ohara Shakai Mondai Kenjujo Zasshi* (Ohara Institute for Social Research) No. 700, pp. 12–19.

Arifianto, Alexander (2009) The securitization of transnational labor migration: The case of Malaysia and Indonesia. *Asian Politics & Policy*, 1(4), pp. 613–630.

Asis, Maruja, and Piper, Nicola (2008) Researching International labor migration in Asia. *The Sociological Quarterly*, 49, pp. 423–444.

Baas, Michiel (2016) Temporary labor migration. In: G. Liu-Farrer and B. Yeoh (eds.), *Routledge Handbook of Asian Migrants*, London and New York: Routledge.

Bélanger, Danièle et al. (2011) From foreign trainees to unauthorized workers: Vietnamese migrant workers in Japan. *Asian and Pacific Migration Journal* 20(1) pp. 31–53.

Chung, Erin Aeran (2010) Workers or residents? Diverging patterns of immigrant incorporation in Korea and Japan. *Pacific Affairs*, 83(4), pp. 675–696.

Constable, Nicole (2009) The commodification of intimacy: Marriage, sex, and reproductive labor. *Annual Review of Anthropology*, 38 pp. 49–64.

Curtain, Richard et al. (2016) *Pacific Possible: Labor Mobility, the Ten Billion Dollar Prize*. Washington D.C: The World Bank.

De Haas, Hein (2005) International migration, remittances and development: Myth and facts. *Third World Quarterly*, 26(8), pp. 1269–1284.

Gaikokujin Kensusei Network (Foreign trainee network) (2006) *Gaikokujin Kenshusei: Jikyu 300yen No Rodosha (Foreign Trainees: Workers with an Hourly Wage of 300 Yen)*. Akashi Shoten.

Geiger, Martin (2013) *Disciplining the Transnational Mobility of People*. New York: Palgrave Macmillan.

Immigration Services Agency, Japan (2019) *Immigration Control and Residency Management*. http://www.moj.go.jp/content/001310186.pdf

International Labor Organization (2015-a) *Countries of Origin and Destination for Migrants in ASEAN*. http://apmigration.ilo.org/resources/ilms-database-for-asean-countries-of-origin-and-destination-for-migrants-in-asean/at_download/file1

International Labor Organization (2015-b) *Bilateral Agreements and Memoranda of Understanding on Migrantion of Low Skilled Workers: A Review*. https://www.ilo.org/wcmsp5/groups/public/---ed_protect/---protrav/---migrant/documents/publication/wcms_413810.pdf

International Labor Organization (2018) *Global Estimates on International Migrant Workers* https://www.ilo.org/wcmsp5/groups/public/—dgreports/—dcomm/—publ/documents/publication/wcms_652001.pdf

Kneebone, Susan (2010) The governance of labor migration in Southeast Asia. *Global Governance*, 16(3), pp. 383–396.

Kretsedemas et al. (2017) Introduction. In: David Brotherton and Philip Kretsedemas (eds.) *Immigration Policy in the Age of Punishment*, New York: Columbia University Press.

Lan, Pei-Chia (2006) *Global Cinderellas: Migrant Domestics and Newly Rich Employers in Taiwan*. Durham: Duke University Press.

Lan, Pei-Chia (2018) Bridging ethnic differences for cultural intimacy: Production of migrant care workers in Japan. *Critical Sociology*, 44(7–8), pp. 1029–1043.

Liu-Farrer, Gracia and Yeoh, Brenda (2018) *Introduction: Asian migrations and mobilities: Continuities, conceptualisations and controversies*. In: Yeoh, Gracia Liu-Farrer and Brenda S.A. Yeoh (eds.), London and New York: Routledge.

Rodriguez, Robyn Margalit (2010) *Migrants for Export: How the Philippine State Brokers Labor to the World*. Minneapolis: University of Minnesota Press.

Ronald, Brown (2012) *East Asian Labor and Employment Law: International and Comparative Context*. Cambridge University.

Silvey, Rachel (2004) Transational domestication: State power and Indonesian migrant Women in Saudi Arabia. *Political Geography*, 3 pp. 245–264.

Silvey, Rachel and Parreñas, Rhacel (2019) Precarity chains: Cycles of domestic worker migration from Southeast Asia to the Middle East. *Journal of Ethnic and Migration Studies* (online) https://doi.org/10.1080/1369183X.2019.1592398.

Song, Jiyeoun (2015) Labor markets, care regimes and foreign care worker policies in East Asia. *Social Policy and Administration*, 49(3), pp. 376–393.

The Japan Times (2016) Lower House Committee Passes Foreign Trainee Bill. (October 21). https://www.japantimes.co.jp/news/2016/10/21/national/social-issues/lower-house-committee-passes-foreign-trainee-bill/

The Japan Times (2019) Probe reveals 759 cases of suspected abuse and 171 deaths of foreign trainees in Japan. March 29. https://www.japantimes.co.jp/news/2019/03/29/national/probe-reveals-759-cases-suspected-abuse-foreign-trainees-japan-171-deaths

The Mainichi Shimbum (2019) Fuhozanryu no gaikokujin jishusei, shakin seoi man-biki (illegal foreign trainees, increased debts, and shoplifting.) January 13. https://mainichi.jp/articles/20190113/k00/00m/040/017000c

The Ministry of Health, Labor and Welfare, Japan (2019) Gaikokujin Koyojokyo no todokede jokyo. (The status of Employment of Foreigners) https://www.mhlw.go.jp/content/11655000/000590309.pdf.

The Ministry of Justice, Japan (2010) *Basic Plan for Immigration Control*. http://www.moj.go.jp/content/000058062.pdf.

The World Bank (2016) *Migration and Development: A Role for the World Bank*. Washington: The World Bank Group. http://pubdocs.worldbank.org/en/468881473870347506/Migration-and-Development-Report-Sept2016.pdf.

The World Bank (2019) *Leveraging Economic Migration for Development: A Briefing for the World Bank Board*. Washington D.C. The World Bank Group.

The World Bank (2020-a.) Personal remittances received (% of GDP) – Vietnam. https://data.worldbank.org/indicator/BX.TRF.PWKR.DT.GD.ZS?locations=VN

The World Bank (2020-b.) Personal remittances received (% of GDP) – Philippines. https://data.worldbank.org/indicator/BX.TRF.PWKR.DT.GD.ZS?locations=PH

Xiang, Biao (2013) *Transational encapsulation: Compulsory returns as a labor-migration control in East Asia*. In: B. Xiang (ed.), (edited by B. Xiang) Durham: Duke University Press.

Xiang, Biao and Lindquist, Johan (2014) Migration infrastructure. *International Migration Review*, 48(1), pp. 122–148.

Yamamoto, Ryoko and Johnson, David (2014) The convergence of control: Immigration and crime in contemporary Japan. *Oxford Handbook of Ethnicity, Crime, and Immigration*. Oxford: Oxford University Press.

Yamanaka, Keiko (1993) New immigration policy and unskilled foreign workers in Japan. *Pacific Affairs*, 66(1), pp. 72–90.

9

DEMOGRAPHY, DEVELOPMENT AND DEMAGOGUES. IS POPULATION GROWTH GOOD OR BAD FOR ECONOMIC DEVELOPMENT?

Sanni Yaya, Helena Yeboah and Ogochukwu Udenigwe

Introduction

The relationship between population growth and economic growth has been debated extensively among economists and demographers. The burning question of whether population growth is good or bad for economic growth has not arrived at a consensus or clear generalization despite several years of debate but proponents of pessimistic and optimistic views on the matter can point to research to support their views. Early discourse on population was initiated by Thomas Malthus who suggested that excessive population growth ought to be controlled to avoid impending (economic) disaster (Peterson 2017). Malthus deemed various types of preventive checks necessary to keep population growth at an adequate level to meet sustenance needs. These preventative checks arguably exist in present-day policies that directly or indirectly influence population trends. For instance, while population related targets are not explicitly mentioned in the sustainable development goals (SDGs), some goals and targets are directly relevant to population trends. The SDG targets in the areas of reproductive health and on universal primary and secondary education will impact population growth (Abel et al. 2016). Proponents of the Malthusian view argue that a fast growing population and high fertility heightens the pressure on natural resources and burdens the economically active population (Mberu and Ezeh 2017). In this scenario, investments in health and education will not be high enough to cover the needs of the population, therefore, resources will be channeled to feeding the growing population thereby impacting future economic growth. Recent studies have reflected this argument. Yao, Kinugasa, and Hamor (2013) concluded that population size impacted economic growth negatively in China. Banerjee (2012) also contended that population growth negatively impacted per capita GDP growth in Australia. In Uganda, Klasen and Lawson (2007) found a negative impact of high population growth on per capita economic growth. Therefore, from the pessimists' viewpoint, countries with higher population growth will record lower economic growth while those with lower population growth are more likely to score higher economic growth rates.

Contrary to the Malthusian argument, other bodies of work have viewed population growth as an opportunity for economic growth. Esther Boserup contradicted the prevailing Malthusian view with claims that population growth catalyzes technological inventions and advancement which spurs economic growth (Marquette 1997). Boserup's views have historical roots in the works of Adam Smith and Karl Marx. In explaining the concept of economy of scale, Adam Smith identified the need for a growing population that will enhance an efficient division of labor. This, as Smith indicated, is key to achieving better return on production (Ucak 2015). Similarly, Marx and Engels explained that in the middle ages, a large population was necessary to allow for division of labor and increase commerce. For Marx, population growth is a consequence of capitalism and a condition for capitalist mode of production (Marx and Engels 1932). Boserup contends that population is an independent factor in developing advanced agricultural technologies which in turn increases productivity. With a focus on sub-Saharan Africa, Boserup argued that economic advancements were hindered by various factors in the region including a historically sparsely population due to the slavery era, a lack of investment in infrastructure dating back to the era of colonization, ineffective governance and Africa's dependence of foreign aid and imports (Marquette 1997).

More recent discourses have posited population growth as a critical component of economic growth. Thomas Piketty, in *Capital in the Twenty-First Century*, contends that the economic growth of a nation "always includes a purely demographic component and a purely economic component" (p.27) (Piketty 2014). Piketty's narrative suggests that high levels of economic growth and economic equality that occurred in the Global North decades after world war II were enhanced by surging birth rates. The sheer bulk of population in China and India is arguably a part of the reason for their current GDP growth rate, which remains among the highest in the world. With the decline in population around the globe, Piketty argues that economic output will be much slower than the norm. Similar discourses agree on the importance of population growth in economic growth and on the detrimental effect of a slow growing population to population growth per capita output while cautioning on the adverse effect of rapid population growth in low-income countries (Peterson 2017). Other recent studies have found statistically significant and positive effects of population growth on economic growth among Common Market for Eastern and Southern Africa (COMESA) member countries (Tumwebaze and Ijjo 2015). The findings show that while population grows, positive factors such as market enlargement, cheap labor, economies of scale and specialization as well as technological progress contribute to increase productivity, which yields greater per capita income.

To the optimist therefore, population growth is among the robust drivers of economic growth (Sethy and Sahoo 2015; Tumwebaze and Ijjo 2015).

Another argument is that the demographic structure of a country provides two channels, fertility rate and mortality rate, and these channels have different impacts on economic growth rate (Mierau and Turnovsky 2013). Population growth resulting from reduction in mortality rates stimulates economic growth while population growth stemming from increase in fertility tend to slow economic growth. Proponents of this argument explain that newborns are considered "resource users" who have little accumulated wealth and so rapid increases in birth rates depletes per capita stock and aggregate savings which tend to reduce economic growth. Working adults on the other hand, are regarded as the "resource creators" and therefore, decreases in mortality rates, particularly adult mortality, will eventually increase economic growth rates.

These unfolding arguments have implications for the African continent where the population continues to grow at a rapid rate. In this chapter, the authors argue that Africa's demographic features do not necessarily spell doom for economic growth in the region. In fact, economic growth can be furthered by a fast-growing population and labor force. We employed extensive literature review; databases searches of the World Bank, UN Department of Economic and Social Affairs; and analysis of relevant secondary data as we sought to analyze the relationship between economic growth and population growth in African countries.

Africa's population growth and trends

Figures 9.1, **9.2** and **9.3** show the global population growth trends, global population estimates, and projections and the continental population share respectively. From 1950–1970, Latin America and the Caribbean region recorded the

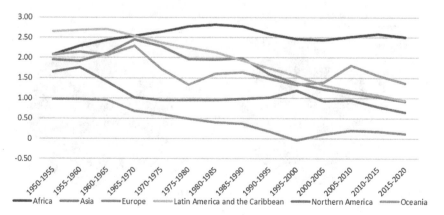

FIGURE 9.1   Global population growth estimates 1950–2020 by regions

Data Source: UN Dept of Economic and Social Affairs, Population Division (2019)

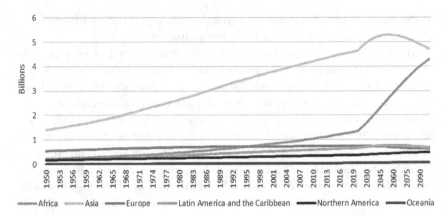

FIGURE 9.2 Global population estimates and projections 1950–2100 by regions

Data Source: UN Dept of Economic and Social Affairs, Population Division (2019)

highest population growth until the period of 1965–1970. Africa on the other hand was at par with Oceania for the 1950–1955 growth estimates but increased afterwards, surpassing the Latin American and the Caribbean region for the 1970–1975 growth estimates, and has become the fastest growing region in the world until current years (see **Figure 9.1**). Africa's demography varies widely. The continent consists of countries like Ethiopia and Egypt with populations of about a 100 million, Nigeria, which recorded almost 200 million in 2019, and others such as Liberia, Mauritania, Botswana and Gambia with population less than 5 million. Regardless of the population sizes of African countries, most of them grow at an average of 2% annually (2015–2020), with few extremities like 3.2% in Democratic Republic of Congo and 0.87 in South Sudan. By 2100,

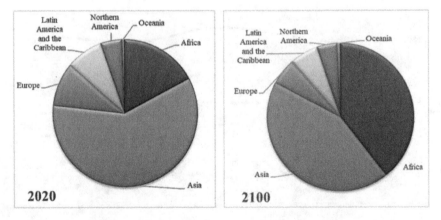

FIGURE 9.3 Continental population share for 2020 and 2100

Data Source: UN Dept of Economic and Social Affairs, Population Division (2019)

Tanzania, Ethiopia, Egypt and Democratic Republic of Congo are expected to join Nigeria to replace Brazil, Russia, Mexico and Bangladesh, which are part of the top ten most populated countries in the world. This means that five out of the ten most populous countries will be found in Africa by 2100 (see **Figures 9.2** and **9.3**). Asia will hit a peak in population around 2055 and then decline but will remain the largest region with projected population share of 43% by 2100, while Africa will experience a surge in population from 2020 to 2100, obtaining 39% of the world population share.

Demographic transition in Africa

Population growth is an important element in the overall economic growth of countries across the world. Evidence shows that whether rapid population growth aids economic growth or negatively impacts it depends on particular country circumstances, such as the population age structure, size of workforce, and birth and death rates of the economy. In other words, it depends on the timing of a country's demographic transition (Headey and Hodge 2009; Huang and Xie 2013; Mierau and Turnovsky 2013). Demographic transition is a process whereby there is an initial phase of high birth and mortality rates and slow population growth. As countries modernize, birth rates remain high, but mortality rates drop leading to increased growth rates. As the transition completes, birth rates will decline resulting in lower population growth.

Countries in Europe experienced a demographic transition with declines in death rates in the nineteenth century and large declines in birth rates occurring in the early twentieth century (Sethy and Sahoo 2015). They transitioned from predominantly rural and agricultural societies with high levels of fertility and mortality to mainly industrial societies with low levels of fertility and mortality rates (Lee and Mason 2006). During this transition, the labor force grew rapidly, and the dependent population reduced, thereby increasing resources for investment in economic development. The process of a demographic transition is currently running its course across African countries as most countries face a unique combination of falling mortality and high fertility. The African continent saw a decline in fertility during the 1970s, however, falling rates of mortality since the 1980s has meant an increase in population. Africa's population currently sees an annual growth rate of 2.5%, a slight decline from a yearly growth rate of 2.6% since the 1990s (Goldstone 2019). Africa's growth rate is expected to decline to 2.3% per year by 2050, but even at that rate, the total population would double every 30 years. This places the African population at a projected 6.2 billion by the year 2100 (Goldstone 2019). Due to the combination of lower mortality and higher fertility, additions to Africa's total population are overwhelmingly young people. Implications of this will be evident in the labor force and the economy. This introduces the concept of demographic dividend whereby young people become productive adults. In exploring the effects of population aging on the economy, Goldstein and

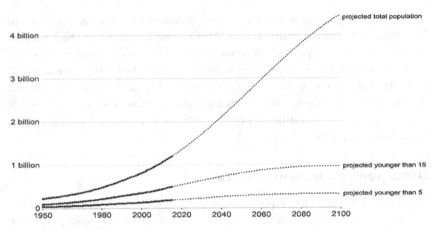

FIGURE 9.4 By age group: the growth of the population to 2100, Africa

Projections from 2016 through to 2100 based on UN median estimates.

Source: UN Population Division (2015 Revision)

Lee (2014) explained that the period in which the share of the population who are of working age grows and the share of the population who are dependents (young and old) shrink gives rise to the "demographic dividend" (p. 199). Although Africa's fertility rate is still higher than other world regions, the difference between its projected population of over 6 billion and young dependent population of about 1 billion, which is the population in the working age, will clearly be higher than other world regions (see **Figure 9.4**). When we therefore shift our focus from population numbers to age structure, Africa can experience positive demographic gains.

Demographic dividend

Demographic dividend as a concept is used in microeconomics to explain the "potential economic surplus resulting from a window of opportunity created by the demographic transition" (Turbat 2017). This demographic window occurs when the working age population increases more than the dependent population which then translates to an increase in the labor force. The surplus here refers to GDP in classical economic terms. To achieve a demographic dividend, the surplus would have to be in excess of what is needed to cater to the needs of the dependents and be well invested and utilized to enhance economic growth and development. Governments tend to invest in economic growth with a strong focus on physical capital (physical infrastructure) and less so on human capital. However, latest research shows that investment in physical, and more importantly human capital holds the key to long term economic growth (Piketty 2014). Under-investment in human capital leads to a workforce that is ill equipped for challenges and opportunities of the future. Even more effective are countries who finance these investments themselves compared to relying on foreign investments. Oftentimes, relying on foreign investments precludes the opportunity for creating economic surplus. As Piketty (2014) explains, wealthy countries who have invested in poorer countries may continue to "own them indefinitely" because as historical evidence suggest, these poorer countries continue to owe the wealthier countries a substantial part of what their citizens produce (Piketty 2014). Asian countries such as China and Japan who are recording high economic growth financed their own investments in human capital of their citizens.

Africa's unique growing population as previously mentioned, indicates an overwhelming addition of young people to the population. This will present opportunities and challenges for African countries. Optimistic views towards population growth will perceive this as beneficial for Africa's labor force and the economy. Africa could use its rising working-age population to yield a demographic dividend. Pessimistic views on population growth argue that African countries face social and economic challenges due to ineffective governance that is unable to cater to the youth's aspirations.

Sub Saharan Africa (SSA) particularly is plagued with poverty, high levels of youth unemployment and unequal access to educational opportunities. Proponents in this camp emphasize the need for expanded educational and health services for youth. Furthermore, the International Monetary Fund (IMF) states that to improve mid-term growth of countries in SSA, an estimated 20 million jobs will need to be created (Turbat 2017; IMF 2019). The concept of demographic dividend makes a link between socio-economic development and population and projects the economic growth that results from young people becoming productive adults. A promising age structure to achieve demographic dividend is one where children 15 years and under fall below 30% of the total population; and people 65 and older fall below 15% of the total population (Goldstone 2019). While demography plays a role in economic growth, investment in a population's education (technological know-how and skills) and efficient government that enforces national commitment to country-specific priority actions are also important recipes for economic growth (Piketty 2014; Canning, Raja and Yazbeck 2015). Indeed, a demographic dividend was evident in East Asian countries during a period of rapid population growth, this was also a period when the dependency ratio was low, meaning that there were more working age people than there were dependent children in the population. These countries saw an increase in economic output per person. Furthermore, there was an increase in investment in secondary/vocational and tertiary education (Goldstone 2019). Many African countries are striving to become emerging economies in the near future. This calls for increased attention to demographic dimensions and how they can impact economic development. The next section is a case study of Ethiopia and Nigeria, countries whose population is matched by a growing economic presence. The case study is undertaken to highlight their prospects for economic growth based on their demographic structure.

Reaping a demographic dividend: The Ethiopian case

The effect of population growth on the economy is evident in Ethiopia, a country on the right path to a population age structure suitable for a demographic dividend (Megquier and Belohlav 2014, Goldstone 2019). As at 2018, Ethiopia had a population of 109.2 million with an annual population growth rate of 2.6% (The World Bank 2020). Furthermore, in 2018, Children between 0–14 years made up an estimated 40% of the population, while adults 60 and older made up 5% of the population (UN Data 2020). With the bulk of Ethiopia's population being between the ages of 14 and 60 years old, Ethiopia's demography presents a hopeful prospect for economic development. It is important to note however, that this does not itself predict a demographic dividend. Megquier and Belohlav (2014) caution that "to benefit from a demographic dividend, countries must first achieve a demographic transition – move from high to low birth and death rates." (p. 2). In recent decades, Ethiopia has made great strides in investing in its health sector. Health improvements in maternal and child health are evident in the rise in modern contraceptive\prevalence among women from 2.9% in 1990 to 35% in 2015 (The World Bank 2020). This has in turn fostered a decline in fertility rate from 7.2 births per woman to 4.5 children per woman. Furthermore, life expectancy is increasing as child mortality remains on a steady decline. The country's life expectancy rose from 47 years in 1990 to 65years in 2015 (The World Bank 2020). These changes have spurred a shift in the country's age structure overtime as seen in the population pyramids below. In 1990 (**Figure 9.5**), the pyramid's broad base represents a large number of younger and dependent children in relation to the working age population. As at 2015 the country showed evidence of a fertility decline with fewer young children in relation to the working age group. In 2030 (**Figure 9.6**), the United Nations projects a proportionately larger working group compared to younger children, based on the assumption of a continued decline in

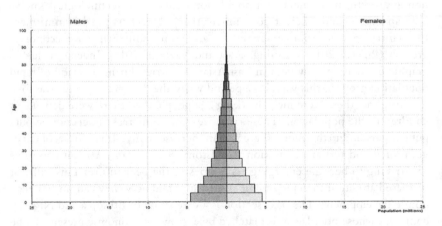

FIGURE 9.5 Population pyramid – Ethiopia 1990

Source: United Nations DESA Population Division. World Population Prospectus (2019)

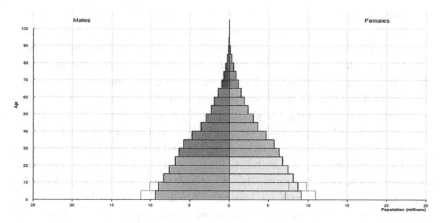

FIGURE 9.6   Population pyramid – Ethiopia 2030

Source: United Nations DESA Population Division. World Population Prospectus (2019)

fertility (Megquier and Belohlav 2014; UN 2019).The demographic changes due to fertility reduction are not sufficient conditions for economic growth. Ethiopia requires a skilled workforce to capitalize on its changing demography and attain a demographic dividend. Investments in human development programs such as education and health is the best strategy to accelerate the economic growth of a country with a surfeit of young workers such as in Ethiopia (Sinding 2009; Megquier and Belohlav 2014). Numerous studies have established a link between attaining higher levels of education and economic growth. Increasing the educational attainment of young people builds a skilled labor force which in turn promotes economic development (Goldstone 2019). Ethiopia has made laudable strides in its education sector overtime; the percentage of primary school age children enrolled in primary school was at 85% in 2015, an improvement from 19% in 1994. However, high school completion rates are relatively poor, with an estimated 9% completion rate (The World Bank 2020).

Reaping a demographic dividend: The Nigerian case

Nigeria's population of 195.8 million as at 2018 makes it the most populous African nation (The World Bank 2019). The country's population grows at an annual rate of 2.6%. The total fertility rate or the average number of children per woman in her lifetime, has decreased overtime. In 1990, a woman in Nigeria had an average of 6.5 children over the course of her lifetime, as at 2017, this number had declined to about 5.5 children per woman. This decline notwithstanding and even assuming a decline in fertility to 3.7 children per woman, Nigeria's population is expected to grow to over 440 million by the year 2050 (The World Bank 2019). High fertility rates in the country have been persistent despite the rise in education which is associated with low fertility (Goldstone 2019). Gross enrollment in primary school has risen drastically from approximately 40% in the 1960s, to 80% as at 2017 (The World Bank 2019). Jimenez and Pate (2017) contend that low-quality education likely negates the gains made by improved levels of education in Nigeria while Goldstone (2019) attributes high fertility rates to a pronatalist culture even in the face of economic modernization. A high dependency rate has persisted in Nigeria for decades. The country's dependency ratio, that is the ratio of dependents (younger than 15 years of age and older than 64) relative to the working age group (15–64), was at 87.3% by 2018. This is a decrease from a 90% dependency ratio in 1990. A higher dependency ratio indicates more people who are not of working age and fewer people who are in the workforce. Implications of a high dependency ratio were shown in a case study on Nigeria. The evidence shows that despite acceleration in the rate of GDP, a high dependency ratio drags on Nigeria's economy because only a small share of the population is working age and more of the population are dependents (Jimenez and Pate 2017).

However, Nigeria's demography is changing. Nigeria is currently experiencing a youth bulge, that is "an increase in the population of the youth group relative to other age groups" (Omoju and Abraham 2014). As of 2018, 44% of Nigeria's population was under age 15 (The World Bank 2019). This means that with continued fertility reduction, the share of the working age population should increase significantly in the near future. In fact, the World Economic Form contend that by 2050 "if employed productively, the working age group can cause Nigeria to reap a demographic dividend" (WEF 2014). These youth will need access to education, healthcare and job opportunities to positively impact Nigeria's economy. The figures below indicate Nigeria's prospects for achieving a demographic dividend based on its demography.

Nigeria's population pyramid shows changes to the country's age structure overtime and projected changes to age structure. The pyramid for 1990 (**Figure 9.7**)

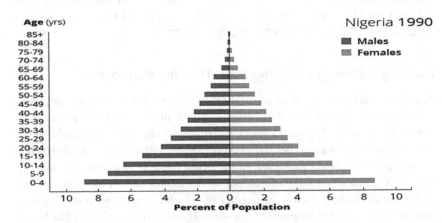

FIGURE 9.7 Population distribution – Nigeria 1990

Source: Population Division of the Department of Economic and Social Affairs of the United Nations Secretariat, World Population Prospects (2012 Revision)

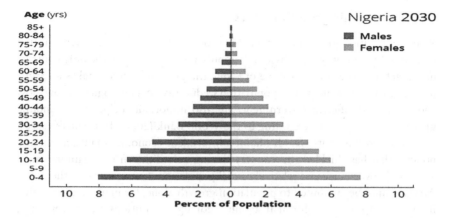

FIGURE 9.8   Population pyramid – Nigeria 2030

Source: Population Division of the Department of Economic and Social Affairs of the United Nations Secretariat, World Population Prospects (2012 Revision)

indicates a high fertility rate and high dependency rates. Changes to Nigeria's age structure are evident in the UN's projections of Nigeria in 2030 (**Figure 9.8**). Fertility rates are expected to decline to 4.9 children per woman thereby narrowing the base. This represents fewer young children in relation to the working age group. Nigeria is projected to further decline in fertility and a larger proportion of working age population who will be able to contribute to rapid economic growth.

Bloom et al. (2010) quantified the economic growth opportunity that would result from Nigeria's demographic transition. The authors project economic growth based on a five-year country level panel data representing a global sample from 1965–2005. In addition to demography, the model considers factors such as human development (education and life expectancy), geographical characteristics and quality of institutions. The study therefore projects economic growth for Nigeria under different demographic and economic scenarios. Under a scenario termed "business as usual" meaning no changes in the current state of affairs, the authors predict that Nigeria will experience a 4% growth in GDP between 2010 and 2030. However, if Nigeria is able to reap demographic dividends along with investments in health, the country's GDP can be expected to increase by 12% by the year 2020 and further increase by 29% by 2030 (Bloom et al. 2010). Nigeria's economy is projected to triple in growth by 2030 compared to only doubling without a demographic dividend. Furthermore, a demographic dividend alongside institutional improvements can lift 34 million people out of poverty by 2030. The need to invest in Nigeria's human capital cannot be overemphasized as this will sustain gains from the country's demography indefinitely. Strong investments in health, education, gender parity and institutions are key policy areas for advancing Nigeria's economy.

Other economic growth factors

While population growth can contribute to Africa's economic growth through demographic transition and dividend, population growth is not the only contributing factor to the rise or decline of economic growth. Piketty makes an argument for the investment in a population's education (technological know-how and skills) and efficient government, are also important recipes for economic growth (Piketty 2014). According to the World Bank (2020), Botswana's success story is the result of its wealth (diamond), prudent economic management and policies that has shielded the country during volatilities in the diamond market, good governance and relatively small population. Growth or lack thereof in Africa cannot be attributed to population growth alone, for many countries that have failed to grow, political instability, corrupt institutions and governance, mismanagement of natural resource, limited investment in health and education and country's geography are key factors (Anyanwu 2014; Zamfir 2016).

Political instability particularly those of violent nature, negatively affect economic growth through the destruction of human and material resources that could have increased the country's economic productivity (Dalyop 2019). Collier (2008) adds that when conflicts happen, investors flee, which affects growth and eventually leads to substantial reduction in economic growth rates. Countries like Democratic Republic of Congo, Liberia, Central African Republic and several others have experienced several political instabilities since independence. Liberia recorded negative economic growth until 1995, in which rates were as low as -51% in 1990 and -35% in 1992, with positive growth rates only recorded in periods just before and after the war (Dalyop 2019). Unlike Botswana that has enjoyed stable political atmosphere since its independence and have continuously recorded relatively higher average economic growth rates, Nigeria since its independence has been plagued with coups, ethnic and religious tensions (CIA 2020). The decline in oil prices in 2016 resulted in a recession in Nigeria, and this recession was exacerbated by militant attacks on oil and gas infrastructure in the Niger Delta region.

Another key factor in a nation's economy is the health of its population. Health and economic growth have a bidirectional relationship. A population's health is one of the most robust driver of economic growth, similarly economic growth contributes to improvements in health outcomes as it reduces malnutrition and increase life expectancy (United Nations Economic Commission for Africa 2019). In the African context specifically, reducing malnutrition will increase African economies by between 3 and 16 percent of GDP per annum (United Nations Economic Commission for Africa 2019). Banerjee and Duflo (2011) add that when unborn babies and children receive better nourishment in the first two years of life, they tend to gain a lifetime income of US $3,269 PPP. Better health outcomes also mean higher rates of school attendance, higher cognitive development and a healthy and productive labor. Inversely, increase in female primary education by one year has been found to reduce infant and child deaths and increase in life expectancy (United Nations Economic Commission for Africa 2019). Furthermore, a reduction in maternal and child mortality rates increases GDP growth (Some et al. 2018, United Nations Economic Commission for Africa 2019). A higher education also fosters economic growth by increasing human capital formation, creating, maintaining, using and sustaining knowledge to further socio-economic growth which translates to higher levels of remuneration, greater tax revenues, improved savings and investment and a more entrepreneurial society (Seetanah and Teeroovengadum 2019).

Conclusion

This chapter discussed the relationship between population growth and economic growth. For many years, this debate has drawn many conclusions with no clear consensus in sight. This discourse was pioneered by Malthus who believed that the higher a country's population grows, the higher the pressure in their natural resources which will eventually lead to underdevelopment and this means the need for policies on population checks to enable the population to meet its sustenance needs. This stance has been challenged by the likes of Piketty and Boserup who point to the importance of population growth in the economic growth and development of economies. To the pessimists who share similar views as Malthus, the arguments are that rapid increasing population is problematic as resources will be channeled to feeding the population while neglecting investments in education and health which are proven factors that lead to increased future economic growth and population wellbeing. This means that for many countries with large population growth rates and population sizes, economic growth rates are expected be stunted or decline. To the optimists on the other hand, population growth is seen as an opportunity to invest in labor force which will eventually lead to higher productivity and economic growth. This group of scholars believe that population growth contributes positively to important factors which yield increased productivity and greater per capita income such as market enlargement, cheap labor, economies of scale, specialization and technological progress. Population growth is therefore a significant positive contributor to economic growth. A third argument is that the demographic structure of the country has two channels, mortality rate and fertility rate, and these channels will have different impact on economic growth.

Decreased mortality rates will increase population growth but will lead to increased economic growth as savings will increase due to increase working adults while the increase in fertility will also increase population growth but will result in lower economic growth since the dependency rate tend to increase.

Africa is the fastest growing region in the world and is projected to house five out of the ten most populous countries in the world representing 39% of the world's population by 2100. With an average annual growth rate of 2% for African countries, many fingers point to its rapid population growth as a huge contributing factor to its poverty and economic growth levels. Evidence shows that when we shift our focus from population numbers towards the age structure, the African region can experience positive impacts of population growth on economic growth. While Africa continues to record higher fertility rates, compared to other regions, by 2100, the difference between its total population and the young dependence population will be around 5 billion and this represents the working population, a figure higher than the projected working population of most of the regions. This promising age structure will enable Africa to achieve demographic dividend which will increase economic growth. Whether a demographic overflow becomes a curse, or a blessing will depend on the well-being of children. This means that for African countries to reap such dividends, there is the need to make the working age more productive, and this can be achieved through; increased education most importantly female education which will empower women, increased access to health care, investment in technology, and increased demand for labor. It is well recognized that good health and higher levels of education are valuable assets to economic growth, in that while 1% increase in secondary education enrollment increased economic growth by 0.309%, proper nourishment of newborns and children for the first two years of their lives leads to a gain of US $3,269 lifetime income. Africa's diversity however requires each country to take country-specific priority actions while ensuring national commitment.

References

Abel, G.J., Barakat, B., Kc, S., and Lutz, W., 2016. Meeting the Sustainable Development Goals leads to lower world population growth, 113 (50).

Anyanwu, J.C., 2014. Factors affecting economic growth in Africa: Are there any lessons from China? *African Development Review*, 26(3), pp. 468–493.

Banerjee, R. (2012). Population Growth and Endogenous Technological Change: Australian Economic Growth in the Long Run. *Economic Record* 88 (281), pp. 214–28.

Banerjee, A. and Duflo, E., 2011. *Poor Economics: A Radical Rethinking of the Way to Fight Global Poverty*. New York, NY: PublicAffairs.

Bloom, D., Finlay, J., Humair, S., Mason, A., Olaniyan, O. and Soyibo, A., 2010. Prospects for Economic Growth in Nigeria : A Demographic Perspective a. *In: IUSSP Seminar on Demographics and Macroeconomic Performance held*. 1–48.

Canning, D., Raja, S. and Yazbeck, A.S., 2015. *Africa's Demographic Transition: Dividend or Disaster? Africa's Demographic Transition: Dividend or Disaster?* Washington, DC: World Bank; and Agence Française de Développement.

CIA, 2020. Africa: Nigeria [online]. *The World Factbook*. Available from: https://www.cia.gov/library/publications/the-world-factbook/geos/ni.html [Accessed 30 Mar 2019].

Collier, P., 2008. *The Bottom Billion: Why the Poorest Countries Are Failing and What Can Be Done about It*. Oxford: Oxford University Press. Oxford: Oxford University Press. https://doi.org/10.1017/S1740022811000118.

Dalyop, G.T., 2019. Political instability and economic growth in Africa. *International Journal of Economic Policy Studies*, 13(1), pp. 217–257.

Goldstein, Joshua R, and Ronald D Lee., 2014. "Economic Inequality ? How Large Are the Effects of Population Aging On." *Vienna Yearbook of Population Research* 12, pp. 193–209.

Goldstone, J.A., 2019. *Africa 2050 : Demographic Truth and Consequences*.

Headey, D.D. and Hodge, A., 2009. The effect of population growth on economic growth: A meta- regression analysis of the macroeconomic literature. *Population and Development Review*, 35(2), pp. 221–248.

Huang, T.H. and Xie, Z., 2013. Population and economic growth: A simultaneous equation perspective. *Applied Economics*, 45(27), pp. 3820–3826.

IMF, 2019. *Sub-Saharan Africa : navigating uncertainty*. Washington, DC.

Jimenez, E. and Pate, M.A., 2017. Reaping a demographic dividend in Africa's largest Country: Nigeria. In: H. Groth and J.F. May, eds. *Africa's Population: In Search of a Demographic Dividend*. Springer International Publishing, 33–51.

Klasen, Stephan, and David Lawson., 2007. "The Impact of Population Growth on Economic Growth and Poverty Reduction in Uganda." No.133. Diskussionsbeiträge. Göttingen.

Lee, R. and Mason, A., 2006. What is the demographic dividend? *Finance & Development*, (September 2006), 1–5.

Marquette, C., 1997. *Turning but not Toppling Malthus: Boserupian Theory on Population and the Environment Relationships*. Bergen, Norway: No. Chr. Michelsen Institute Development Studies and Human Rights.

Marx, K. and Engels, 1932. A Critique of The German Ideology. In: *A Critique of The German Ideology*.

Mberu, B.U. and Ezeh, A.C., 2017. World population estimates and projections 1800– 2100. *African Population Studies*, 31(2), pp. 3833–3844.

Megquier, S. and Belohlav, K., 2014. *Ethiopia's key: young people and the demographic dividend*. Washington, DC.

Mierau, J.O. and Turnovsky, S.J., 2013. Demography, growth, and inequality. *Economic Theory*, 55 pp. 29–68.

Omoju, O.E. and Abraham, T.W., 2014. Youth bulge and demographic dividend in Nigeria, 27 (Mar), 352–360.

Peterson, E.W.F., 2017. The Role of Population in Economic Growth.

Piketty, T., 2014. Part one: Income and Capital. In: *Capital in the Twenty-First Century*. Cambridge, Massachusetts. London, England: The Belknap Press of Harvard University Press, 1–112.

Seetanah, B. and Teeroovengadum, V., 2019. Does higher education matter in African economic growth ? Evidence from a PVAR approach. *Policy Reviews in Higher Education*, 3(2), pp. 125–143.

Sethy, S.K. and Sahoo, H., 2015. Investing the relationship between population and economic growth: An analytical study of India. *Indian Journal of Economics & Business*, 14(2), pp. 269–288.

Sinding, S.W., 2009. Population, poverty and economic development. *Philosophical Transactions of the Royal Society B: Biological Sciences*.

Some, J., Pasali, S. and Kabone, M., 2018. *No Healthcare and Economic Growth in Africa*.

The World Bank, 2019. Nigeria Data [online]. *The World Bank Data*. Available from: https://data.worldbank.org/country/nigeria?view=chart.

The World Bank, 2020. Ethiopia | Data [online]. *The World Bank Data*. Available from: https://data.worldbank.org/country/ethiopia?most_recent_year_desc=false [Accessed 16 Jan 2020].

Tumwebaze, H.K. and Ijjo, A.T., 2015. Regional economic integration and economic growth in the COMESA region, 1980–2010. *African Development Review*, 27(1).

Turbat, V., 2017. The Demographic Dividend: A Potential Surplus Generated by a Demographic Transition. In: *Africa's Population : In Search of a Demographic Dividend*. 181–195.

Ucak, A., 2015. Adam Smith: The inspirer of modern growth theories. *Procedia - Social and Behavioral Sciences*, 195(284), pp. 663–672.

UN, 2019. Word Population Prospects 2019 [online]. *World Population Prospects - Population Division*. Available from: https://population.un.org/wpp/Graphs/ DemographicProfiles/Pyramid/231 [Accessed 20 Jan 2020].

UN Data, 2020. Ethiopia-Social Indicators [online]. *Ethiopia-Country Profiles*. Available from: http://data.un.org/en/iso/et.html [Accessed 15 Jan 2020].

United Nations Economic Commission for Africa, 2019. *Healthcare and Economic growth in Africa*. Addis Ababa, Ethiopia.

WEF, 2014. *Prospects for Reaping a Demographic Dividend in Nigeria*. Gevenva, Switzerland.

Yao, Wanjun, Tomoko Kinugasa, and Shigeyuki Hamor. 2013. "An Empirical Analysis of the Relationship between Economic Development and Population Growth in China." *Applied Economics, Taylor & Francis Journals* 45 (33): 4651–61.

Zamfir, I., 2016. *Africa's economic growth: Taking off or slowing down?*

10

SHARING THE PIE? THE FOURTH INDUSTRIAL REVOLUTION AND THE SHARING/PLATFORM ECONOMY

Distributive justice implications

Rasul Bakhsh Rais

Introduction

The modern era we live in today has witnessed rapid industrial and technological transformations that in consequence have greatly reshaped the economies, life-styles and the larger social world within which we interact. The way we live, the way we think and the things we consume—some of the aspects of culture—are path dependent. Whether this path is taken as a conscious and planned decision or it is a spin-off effect of new forms of economy and technological innovation, its consequences for human behaviour, political ideas and politics can be immense. I take a liberalist position on politics being instrument to what we dream as our future, which resources we mobilize to achieve our objectives and how we distribute the rewards fairly and justly. Therefore, the modern economic systems we work in and the technologies we consume are not an accident of history, but an outcome of decisions the major players and actors shaping the global and national economies take. Since politics is a purposive act, those interested in political economy would investigate questions of ends and whether or not, and how the rewards are distributed within the society and across the borders in the larger global world.

Globalization, no matter what perspective you take on it, and the technologies driving it are not independent of the intentions and interests of the powerful players at the world stage. Investments in research and development, science and technology both in public and private sectors are manifestations of national purposes and the profit motives of the corporations. Underneath all of this is the critical factor of dominant ideas and ideology—capitalism at large, and the neo-liberal order with its focus on the market forces, capital, privatization, limited government and lesser regulation. The "triumph" of liberalism and its celebration in the dominant policy making circles in the western world and intellectual hegemony in the academia seems to have squeezed spaces for alternative view and thinking. With technological innovation, efficiency in production and introduction of new products, even altering the consumer behaviour in many significant ways to market new products and services, the value of people and importance of distributive justice have diminished. Rather, they seem to be missing in both the intellectual and policy conversation.

The fourth industrial revolution that seems to be shaping up promises new economy, offers new business models by presenting smarter tools and devices carries with it many opportunities and challenges with it. Rather, its impact on everything from government to people, businesses to market, culture to national and global security is likely to be at a scale and in scope never seen before; on a greater, and even unprecedented level than the previous three revolutions. Since the first industrial revolution and capitalist mode of production, efficiency, cost reduction and integration of supply and demand in the marketplace have been the driving ideas behind the industrial economies. The emerging fourth industrial revolution rests on the same concepts but with a significant difference in speed, wider and rapid spread beyond the developed world and the impact it is going to have on every aspect of economy, politics and society.

The technological revolution has the potential to cause systemic and structural changes that require "comprehensive and integrated" response from all stakeholders, the governments, businesses, academia the civil society and international institutions (Schwab 2016). Inequality and inequity associated with the market economy is not something new, and we know how sentiments of unfairness generate troubles when larger and critical sections of the population become disillusioned. Compared with any other eras of modern history, disruptions in the labour market around the world are likely to be more, extensive and intensive. Joblessness, work force dislocations due to the new dynamics, sinking of old and rising of new business and varying impact on income levels are some of the trends we have already observed resulting from the fourth revolution. As the revolution is full of promise and opportunities the challenges it presents are monumental. On the one hand, we may see rising income level, new job opportunities and innovative business enterprises, on the other bigger gaps in levels of income The basic argument of this chapter is that creation of wealth through the combination of new technologies, capital and innovation may generate new levels of prosperity, but the gains may not be distributed fairly and equitably within the nations and across the globe. The disparities of income may be wider, losers in greater numbers and the beneficiaries in minority, but capable of raking in rewards at much larger scale than in the past. It is quite possible that the combination of factors of production and the traditional role of the capital may change. Fuelled by the energy of fourth revolution, the dynamics of the economies may promise greater rewards for talent, expert and scientific entrepreneurs than any other factor. The issue of distributive justice may gain fresh impetus, steam new social movements and imperil stability of the existing social and political orders.

Firstly, the chapter explains the fourth industrial revolution and the salient trends unfolding for the past many years. Secondly, it examines several manifestations of the sharing or platform economy. Thirdly, it analyses the disruptive potential of the revolution and the economy it is spanning and why distributive justice—sharing the pie is an important and relevant issue in mitigating the negative effects of the revolution. Finally, political challenges emanating from the neo-liberal order and the winner-takes-all attitude are discussed. The kind of politics the leaders and the nations choose will make social and economic transitions smooth or troublesome. While the acceptance of distributive justice as a policy tool may aid the efforts to accept and absorb new innovations and reconcile conflicting interests and resistance to it may generate political storms, social? and invite trouble from the critical sections of the populations entertaining a feeling of having been left out.

The fourth industrial revolution

In the past few centuries, we have seen one industrial revolution leading to another one. In the first one, water and steam replaced human and animal labour. That is what started the first wave of mechanisation and production of industrial goods and a major transformation in the transport system. The second revolution rested on the electric power that entirely changed the way we lived and things we produced and consumed. Third is the digital transformation of everything from factories, offices to every means of communications (Pouspourika 2019). People of my generation have seen this happening from introduction of first generation of apple computers of merely12-k memory using floppy disks to quantum computing, mobile phones and fast transition to smart phones. Having lived through this continuing revolution over the past forty years, since our graduate school years, the technology firms, innovators and the scientific talent has never stopped giving us pleasant surprises. In a country like Pakistan where I live and work, owning a landline without direct dialling to next town was a luxury just thirty years back. Now there are 165 million out of a population of 220 million have use cell phone and 75 million have access to Internet.[1] I am referring to it because this means of communication has changed much of our social, economic political and information world. It is difficult to single out one critical factor shaping the new revolution. It comprises many elements of several technologies, industries and streams of scientific knowledge, often blurring conventional disciplinary lines between them. How can we think of global reach of the cell phone, the Internet, messaging services without satellites, fast computers, and data storing services?

Let us briefly look at the some of the facets of the fourth revolution over the horizon, becoming more visible and larger and larger by the day and covering more and more areas of our professional and daily life. During the COVID-19 pandemic closing down the world economies, the cities, factories, offices and the educational institutions, amazingly the communication technologies contributed greatly to continue meetings within and across the nations, delivering courses on Zoom and other similar platforms, and ordering essential food, medicine and other goods online. Gladly this transformation is not confined to the developed world. The developing countries that have vested in 3-G and 4-G communication technologies have reaped the benefits. As the world is striving to cope with the pandemic, trying to balance between individual safety and urgency to open up essential elements of the economy, tens of billions are being invested in finding the treatment and develop a vaccine. Time is an important variable to save millions of lives. We will find the vaccine, and sooner than later by shortening the period required in understanding the genetic makeup of the virus, its origin and cause, development and testing. That will be possible by only application of Artificial Intelligence, quantum computing, and instant integration and communication among the members of scientific community working on the project in different countries. Work of years is being squeezed into months by scaling up the process using high-speed computers (Knvaul Sheikh 2020).

Throughout the world, online shopping, telemedicine, home delivery of food, medicines or any other goods has accelerated. No one considers it a luxury but very essential to the modern urban life. Now billions of people around the world have access to smart phones making interpersonal, commercial, governmental and scientific communications easy, cheaper and efficient. Actually, it has pulled down many of the barriers barring a few authoritarian states in accessing information. Access to global sources of knowledge for the academia, students, businesses and general public are immense. It looks like living around a library with thousands of online journals, newspapers and direct access to libraries around the world.

Some other marvels are equally notable, like Internet of things, energy storages, databanks, nanotechnology, biotechnology, autonomous vehicles and many amazingly rapid developments in the field of 3-D printing and material sciences (Schwab 2018). The impact of the technological transformations we are witnessing may go beyond what we have seen so far. Our societies, economies, lifestyles, workplaces, production and distribution of consumer goods, and even the architecture of the urban spaces is going to change. We are not estimating the true potential power and influence of the social media platforms on cultivation and spread of ideas from political to religious, ethnic to cultural in globalised forms when the barrier of the state are bound to weaken. Two things noticeable even at the early stages of the fourth industrial revolution are, it cuts across the conventional boundaries between the "physical, digital, and the biological" worlds. Ensuing from this, it is "fusing advances in artificial intelligence, robotics, the Internet of Things, 3D printing, genetic engineering, quantum computing, and other technologies" (McGinnis 2018). The disruptive potential of platform economy is fast bringing forth innovative ideas, and novel ways of marketing goods and services, shifting consumer interest away from the traditional services and outlets to more comfortable and economic means. What we can foresee is a continual flow of new service providers in the consumer market with new tools using the marketing power of the digital media platforms to change the consumer behaviour. The quality of services, ease of doing business, affordable pricing and consumer satisfaction are some of the ideas new entrepreneurs are using to compete with traditional, existing industries and businesses. Another defining trend of the revolution is building business alliances within the territorial boundaries of the state and entering into collaboration with similar international ventures. This benefits all by pooling information resources, calibrating business models by using same methods of data collection and use, sharing capital and technology. This is bound to make the marketplace more vibrant, competitive where better quality of products and services will retain the trust and loyalty of the consumer. However, we cannot escape the disruptive effects on jobs, manufacturing, distribution networks, businesses, and more importantly on how the wealth by new businesses is shared by all the stakeholders fairly and justly. Let us understand first what is the sharing economy and what potential it hast to create new sources of wealth and diminish the power and capacity of some of the traditional modes of production, marketing and distribution.

The platform economy

According to Schwab, "technology enabled platforms that combine both demand and supply to disrupt existing industry structures, as those we see within the 'sharing' or 'demand economy'" is one of the most salient features of the fourth revolution (Schwab 2016). This captures essence of the concept. However, in an emerging and fast evolving new form of economy, we need to further spell out various dimensions of the sharing economy. Essentially, new, and some old, entrepreneurs are increasingly exploring and applying new technology platforms to integrate producers, consumers, and markets in order to facilitate access to under-utilised "access of goods or services between two or more parties" (Miller 2019). Within less than a decade, the new companies entering the sharing economies have proliferated, expanded quickly and have achieved remarkable success in the market place. There are three types of individuals that seem to be showing great interest in the sharing economy; rather driving it. First, the people who have free time, semi-retired or would like to freelance preferring to make their own time, and choose something that reflects their interest, level of income expectation and match their skills. Second, the price conscious consumers shopping for ease and lower prices. Third, the investors that understand the potential of innovative businesses and the role of platform technologies in harnessing idle or under-utilised assets and workers.

One of the major reasons for the expansion of the sharing economy, businesses and companies is that they "offer competitive prices to consumers in part by acting as labour brokers, rather than as employers who offer costly employee benefits" (NYT 2014). There are no inventories, no overheads and no massive payrolls. It is about making right connections between asset holders, workers and the consumers. Everyone involved in this triangle benefits. Some of the examples are ride-hailing services, chore marketplace, freelancers, creative and professional services providers, delivery services platforms, grocery delivery services, which during the Covid-19 have assumed much greater importance than ever before. What we have observed in Pakistan during this crisis is that competition from the emerging small-scale services, has forced the conventional grocery stores to start their own delivery services. According to one source the venture capital investment in these businesses was $2.19 billion in 2014. Uber claimed the largest share of venture finance then, which was estimated as $1.5 billion. Even TaskRabbit, a chore service company received $38 million (NYT 2014). Since then, they have all grown much bigger. Many other interesting ideas have caught imagination of people in the countryside, like land sharing. Private property owners have started attracting people looking for alternative sites for camping in the rural areas, or in private gardens in the urban centre to beat high cost of hotel accommodation. Privacy, back to nature and lower prices are contributing to the rise of ventures like Hipcamp in California and Gamping in France (Hardy 2015). Since many of the ventures, and in so many areas, are private, and evolving fast, it is difficult to cover all the areas of their operations or estimate their net collective worth.

There is an agreement among the economists that sharing economy is the fast growing in every corner of the world from small scale to large scale, national to regional and international. Within about nine years, between 2010 and 2019, venture capital firms invested $23 billion in sharing companies (Miller 2019). According to the same source, "Airbnb ($31 billion) and Uber ($72billion) have a combined $103 billion market cap which would rank them as the 38th wealthiest country in the world".

Let us briefly have look at some of the players in the sharing economy and how they seem to be causing major shifts both in the business patterns as well as in the consumer attitudes. The important thing to note is that workers, assets, consumers have all exited, and were all part of the conventional economy. The communication technology is the critical factor that has changed old transactional relationships by creating and connecting new stakeholders with the well-known incentives of shared resources, shared marketplace and shared benefits.

Transportation

Uber was perhaps the first in the market, but many others like Didi and Lyft followed the lead. In the Middle East and South Asia Creem, a new brand has copied the same model, but then taken over by the Uber. Local start-ups in this sector are too many in each country to document, which are offering services to the lower-end consumers by offer motorbikes and motor rickshaws as an alternative to the public transport. Found in San Francisco in 2009, Uber operates in 900 cities around the world. Its valuation in 2018 was $72 billion and revenues in that year $11.3 billion. It employs 3.9 million drivers (Iqbal 2020). Uber's closest competitors is a major ridesharing service, Didi, a Chinese company, which according to Forbes achieved a valuation of $50 billion and aims at raising it to $80 billion, if that happens, would surpass that of Uber (Forbes 2018).

Lyft is emerging as a close rival and competitor to the much resourceful, richer and well-established global ridesharing brand, Uber. The Lyft is building its reputation as a more transparent and secure company to better compete with Uber. So much was the trust of a large variety of major global players from General Motors and Fidelity to Rakuten, a Japanese e-commerce company and Saudi Kingdom Holding Company. It raised $5 billion from over one hundred investors (Bort 2019). In May 2019, its valuation was $15.1 billion and the company aims to push it up to $20 to 25 billion (Forbes 2019). It is very novel model of business that brings revenues to these companies just for owning and operating the technology platforms without any investment in assets or paying any salary to the drivers. It is altogether a new economic ball game. Having downloaded the application of any ridesharing service a consumer using smart phone device books a ride and at the destination pays cash to the driver or uses credit card. Out of this deal between the driver and the rider, Uber, Lyft or Didi takes roughly 25-27 percent from each ride. With billions of rides annually one can imagine the wealth that is being generated and how investors in these companies are profiting from the destruction of traditional taxi services. While the ride-hailing companies tout their business as a "win-win" for all stakeholders—the consumer, investors and drivers—the unions of the taxi services have been protesting and seeking protection of their livelihood from the city managers.

The obituary of traditional taxis is written all over the globe. The ridesharing companies have rapidly replaced the conventional taxis as a popular means of urban transportation in metropolitan cities around the world. Instant location data provided by Google maps, spread of smart phones and easy to obtain ridesharing are some of the factors that have made the difference. While it may be a bonanza for venture capital investors and technology companies, the cab drivers all around the world have increasingly lost their investments, business and employment. Attempts to cap the number of ride-hailing cars entering annually in places like New York have eventually failed. The taxi unions rightly attempted to make a "social justice" issue out of the exponential expansion of ride-hailing companies, and for a while the politicians paid some attention to their cries. Sadly, they have lost out. Interestingly, Uber and Lyft contested their case in the media, before public and the halls of power as "civil rights" issue, and they finally won (Mays 2018). After losing investments earnings, some of the taxi drivers have ended their lives out of financial distress (Fitzsimmons 2018).

Consumer goods

According to consumer marketing and research three factors matter in consumers' decisions in accessing a traditional outlet or shared economy platform. These are affordability, convenience and efficiency (Dillon 2016). Consumer surveys suggest that people switch fast to shared-economy for convenience and lower cost reasons (PWC 2015). The shared-based products and varieties are fast taking over a significant portion of the market. One of the pioneers in the consumer market is eBay that started its modest operation as online auctioneer of goods in 1995. It has become a multibillion dollars international corporations operating in 33 countries worldwide. In 2019-2020, its revenue was $10.7 billion (Celment 2020). Its earnings are primarily from the fees its collects from sellers and buyers that use its platform. It has a negative list of goods that cannot be traded, but other than anything can be sold and bought. The sellers post images of articles they wish to sell with a description of their conditions, prices and contact details. They mail the goods directly to the buyers. In this peer-to-peer transaction, eBay collects its fee. The net value calculated by the stock prices is estimated to be $43 billion in 2020; much lower than it was couple of years back (Okenwa 2020).

Today, eBay is not the only a player in the consumer market, offering interface between the buyers and sellers; Etsy and Rent the Runway are its second and third bigger competitors but more are likely to enter the consumer goods shared economy. In June 2020, the market value of Etsy was $11.32 billion and its stock price higher that the eBay (Macrotrends 2020). The Rent the Runway is one of the most innovative of businesses; it rents designer dresses to women. Its value in 2019 was estimated to be $1 billion. It's aiming to become an "Amazon Prime of Rentals" by expanding renting to goods like quilts, pillows and bedding, the daily use items (Maheshwari 2019).

Professional services

Today, in every society, there is a large pool of workers with skills, experience, and diplomas certifying some expertise to perform certain services. The examples can be masons, carpenters, plumbers or accountants. We have seen that some of them would like to offer their services as part-timers to augment their incomes. It is utilisation of free available time according to one's convenience. In the modern labour market, these workers are known as freelancers. As people would like to hire contractors for work after making sure they are getting someone who has requisite skills, experience, and credibility. Fiverr, Upwork and TaskRabbit are new digital economy brands that trade human expertise the way material goods traded in the marketplace. Fiverr alone provides freelance skilled labour in 300 categories. The TaskRabbit has very interesting ideas to offer to consumers from delivering groceries to help move from one place to another to home repairs fixing faucets and running errands (Taskrabbit 2020). The trend in hiring labour for a short time with the per hour wage announced and with the guarantee of the company appears to be growing. In 2014, about 25% of the American labour force was freelancing, which was estimated to double by the end of the decade. Flexibility of both the service provider as well as that of the consumer has been the driving force behind the expansion of online professional services.

How these services fare on equality, social justice and non-discrimination. If all factors are equal, a white worker is likely to be hired more often than a black or any other person of colour. There is also an issue of feedback that consumers give to the company after a task has been performed. Studies conducted so far suggest, blacks in the United States don't get as high evaluation as the white workers hired for the same job. The profiling of the workers on website doesn't make many people comfortable in societies where race and racial prejudices are issues (Hannak et al. 2017).

Healthcare

The pandemic, Covid-19, has greatly raised the importance of collaborative consumption in the health sector. In many countries around the world, especially the developing ones, as the outdoor patient services closed down due to fear of virus spread, the public as well as private sector hospitals began offering consultation online. It has proved to be an effective means of seeking medical advice and getting a prescription for any ailment without the trouble of queuing up in a hospital (DAWN 2020). Well before the global virus crisis of 2020, doctors around the world had set up online consultation services, which have been cheaper, efficient and only a click away. The changes in the healthcare provision by established and emerging digital start-ups would leave far-reaching effects on this sector. Its potential is in sharing medical equipment and idle facilities, serving people in the countryside, physically impaired persons who cannot travel to the hospital by using video links. This is what Doctors on Demand, American Well and Cohealo are doing besides telemedicine consultation that includes giving an option to patients for selecting their own doctors and nurses (Abelson 2016).

Telemedicine is an emerging sector of the sharing economy in every country and at the global level. In 2018, this sector generated revenues worth $12.5 billion in the United States alone (American Well 2019). Given the technological advancement in the developed world, this sector appears to be growing fast. However, the real potential of distance medical care has yet to be realised as it is in the early stages of development. In the developing world where all types of inequalities and inequities exist, access to healthcare is poor and expensive. There appears to be an alternative in building telemedicine networks to improve access to health facilities. The World Health Organization is running many of such facilities around the world, which many countries have started emulating to reduce the effects of curable diseases, like blindness and infant mortality rates (Wootton et al. 2012).

The investors see a big opportunity in ventures like telemedicine and collaborations among the doctors and conventional hospitals where expensive medical equipment and space remain unused. Multiple healthcare facilities can use the same equipment by using sharing technology and data on capacity and under use equipment and facilities. The most important development is rapid shift from "traditional hospital settings to consumer-focused sites of cares", in geographical proximity to patient's home, or at home that is expected to reduce costs of healthcare (Singhal and Carlton 2019). Many of the technologies of the fourth revolution are expected to bring about a revolution in the conventional healthcare system with a focus on sharing logistics, home-based, personalised treatment and wider choice for the patients in accessing facilities and selecting their own doctors and nurses.

Distributive justice

Having discussed the fourth industrial revolution and some of the emerging facets of the sharing economy, it is pertinent to raise some questions about the distributive effects of the economic and technological progress the mankind appears to have achieved. We cannot emphasis the point more that changes we are witnessing in the technology-driven world economy of the fourth industrial revolution are "not merely a series of incremental technological advancements. It is an upheaval—a dramatic and wide-ranging shift in the way that value is created, exchanged, and distributed across individuals, organizations, and entire economies" (Schwab 2018). As we have seen in the previous three revolutions in our history of technological and industrial progress, inequities and inequalities have widened. Never have technologies and modes of production at any time in history been neutral or independent variables. Using the Marxist analytical term, the "relationships of production" matters. It means who owns them, which particular classes or layers of society are more likely to acquire technological knowledge, managerial and entrepreneurial skills are very important factors in the modern age. In the new age economy, talent that drives the technological progress and the capital, which has remained the critical element in every industrial revolution have claimed much larger share of the wealth that has been generated. Above all, it is a structural factor, a pattern or set of principles under which things are lawfully acquired, held and transferred. Robert Nozick calls this "principle of justice in acquisition" or holding things as one's lawful entitlement (Nozick 1973, p. 51). In modern times, the economic and political systems, the ideology behind them, the class structure that is reflected in every aspect of the economy and society has established the principles of just and rightful ways of acquiring, transferring and holding things. While the socialists would like to apply "from each one according to his capability" and to "each one according to his needs" as a just principle of distributive justice, the liberal theorists rest their definition on individual choice in a free society, but they don't seem to be oblivious to the results of distribution. According to F.A. Hayek, "our objection is against all attempts to impress upon society a deliberately chosen pattern of distribution, whether it be an order of equality or of inequality" (Hayek 1972, p. 87). What follows from freedom is then "to each according to how much he benefits others who have the resources for benefitting those who benefit them" (Hayek 1972, p. 87). Nozick refines this idea by saying "from each as they choose, to each as they are chosen" (Nozick 1973, 56). Much of the writings of past two centuries fall essentially within the conceptions of liberty, pluralism, freedom of choice, and more importantly the utilitarian principles. In contrast, John Rawls in his classic *Theory of Justice* makes an argument for "the socially just distribution of goods in a society", by invoking the principle of cooperation and mutual aid in the production of wealth. The main idea of this theory is that free, rational persons motivated by personal gains and interests would accept an "original position of equality" as a defining and fundamental principle of their being a member of a single community (Rawls 1971).

Amartya Sen appears to be critical of Rawls's argument that fair procedures would mean justice, as it would be difficult to establish on account of the fact there can be many principles of fairness. Sen on the other hand, makes a case for "reasoned choice of just policy" by insisting "for an adequate understanding of the demands of justice, the needs of social organizations and institutions and its satisfactory making of public policies" (Roger 2010). Sen is more concerned with the resulting inequality from freedom of choice. He uses the conception of democracy as "discussion" more than the idea of democracy as "elections" to make the point that while maintaining the principles of pluralism and rational choice, we can use the instrument of public policy and institutions to address larger issues of inequality in the society.

The challenge we face is in the ideological template of neo-liberalism, the systemic factors. We shouldn't expect different results from the fourth revolution in terms of equitable and fair distribution of gains that the national and global economies are going to make, if the quest for justice, fairness and equality and the voices in the academia and politics for distributive justice remain weak or muted.

What are our rights and duties, how we are to be rewarded for work, how we treat one another as members of the same community, what we owe to the political authority and what obligations it has towards citizens are some of the perennial questions of political theory. These are some of the questions that define the concept of social justice. Equality before law, fairness, non-discrimination and equitable distribution of rewards are some of the ideas that have never lost importance and relevance in any age. They are more important when the inequality worldwide and within the nations has significantly increased at a time when our societies have achieved an unprecedented affluence. According to "The World Social Report 2020" of the World Bank "the richest one percent are the big winners in the changing global economy, while at the other end of the scale, the bottom 40 per cent earned less than a quarter of income in all countries surveyed" (UN 2020, n.p.). While global inequality has slightly declined due to growth in two populous countries, China and India, inequality within nations has been on the rise. Regional disparities among countries have stayed on the old pattern. For instance, income level in North America are about sixteen times higher than the in most developed countries, and some middle-income countries than Sub-Sahara Africa (Bourguignon 2016).

The Neo-liberal economic policy reforms in the developing world have opened markets for investment and shifting of industries and capital in a big way. Thus, we see the economies globalizing and connected, but the inequalities have also taken stronger roots than ever before. The effects of inequality are well known on disparities in healthcare, education, life expectancy and infant mortality rate. In countries where rich have become richer and poor poorer, we have generally witnessed people suffering from poverty-trap, unable to change their economic and social circumstances. Overall, impact of inequality is negative on economic growth and investment in developing the human capital. The evidence we have so far shows that the fourth industrial revolution sweeping the world has yet to make a dent in the old ways economic benefits of progress are distributed in the society.

With emergence of sharing economy in so many diverse unconventional ways, conflicts between those having stakes in the old business models and the new ones are unavoidable. One of such cases is the Airbnb listings rooms for rents for a short period worldwide and the hotel industry. Besides, the question is how this new business can be regulated in the same way as hotel industry pays local and national taxes and where investors have made huge investments. Operating within wide legal loopholes, beating the existing regulatory regime raises serious questions of fairness and equity. It may not replace the hotel industry but is certainly claiming a share in the market by means and practices that fall out of the net of regulators. It also impacts gravely rent market in major cities around the world, in some cases, property owners have converted apartments into rooms for rents for short-term renting with the Airbnb that may bring more revenues than long-term renting (NYT 2014).

There are many well-known legal and economic means of providing a plain-levelled-field to all actors in the marketplace and pursuing progressive policies of ensuring a minimum decent standard of living for the poor. It can be done by welfare policies and progressive taxation to redistribute wealth. Legal mechanism would include fair regulation to prevent cartels, price fixing and monopolies.

There are major concerns about how the wealth produced through the sharing economy can get concentrated in the developed world, and within the developing world a particular class of people can benefit from it more than others. Many of the sharing or platform companies and business we have discussed above are operating from their headquarters in the Western world, of course providing economic opportunity, growth and employment in other parts of the world. For owning and managing the technology platforms they get a share from each transaction, causing constant flow of capital to the rich countries, which amounts to billions of dollars each year. So far, they have remained the major players in the local markets and attempted to gobble up any emerging local competitor. Powered by technological edge, innovation, constant flow of capital and an international brand, they dominate the markets. Their size and global presence gives them unfair advantage to any new competitor. The legal and regulatory regimes, elite corruption and class structure of the society make it very difficult to ensure fairness even by the standards of neo-liberal economic policy framework.

There is another important dimension we need to look at from the point of view of social justice. This is the plight of taxi owners and drivers in many parts of the developing world who have lost out to the Uber, Creem and similar ridesharing services. Such drivers are generally illiterate, poor and heavily dependent on daily income from running taxies to feed their families. They can now operate but find the business unsustainable. They can do so on the margins of the cities and towns where the ridesharing companies have captured the central sites and the relatively affluent sections of the urban travellers. The middle and upper middle classes that see convenience and hope to save money by entering the collaborative consumption economy are sold out on the sharing economy. That is what happens in the market economy when each consumer acts to maximize his gains more rather than looking at the bigger picture of the society and what negative consequence it might have for other classes.

Not everyone looks at the downside of the sharing economy or issues of loss of jobs and dislocations. There is a sort of new optimism among neoliberal economists and commentators about the sharing platforms and the distributive effects of gains. They see it in the light of "positive externality", meaning that sharing of existing, unused assets, space and goods can generate income for all participating in it, or save costs for those using them. The "externality" is that of the technological platform offering links, data and managing transactions. A commentator in *The Guardian* captures this idea by saying that "the utility is transparent and the benefits distributed more transparently" (Mason 2015). He seems to be entertaining unbounded optimism about potential benefits of sharing economy on social justice. He argues "Today, if you wanted to re-order the economy to deliver participation and choice alongside social justice, it is the sharing models you would start from" (Mason 2015).

Conclusion

Surely, the sharing platforms offer wide range of choices to the consumers. The technology entrepreneurs may find more job opportunities, and workers would like to use spare time for some extra income. One cannot discount the value one can add to vacant and unused asset. We all see the heralding of a new economy that appears to integrate demand, supply, consumers, technology and investors, creating a wide participator web or economic relationships. The revolution in the communication and digital technologies has made all this possible. Having said this, the fact is that this economy is squeezing a bit of more value out of the existing facilities and assets, skills and available time than investing more in them.

The bigger question is, are all the gains put together enough to justify harm to workers, communities, professions, businesses, and consumers? Are these models of sharing economy environmentally sustainable, as we fear cheaper fares may bring out more cars on the streets, turn people away from the public transport system? What impact travel and stay in bigger cities in cheaper accommodation through the Airbnb platform mean for crowed cities and tourist spots? In this early stage of unfolding economic revolution, we don't find clear, evidence-based answers to all these questions. But it is important to raise these questions for future research and for the deliberations of the policymakers.

At the moment, it seems an optimistic outlook of the sharing economy prevails, particularly among the consumers, as they find sharing platforms convenient and a cheaper alternative to conventional means of transport, buying goods, hiring services and accessing healthcare facilities. It seems the sharing economy companies have gained the trust of the consumers, asset owners and venture investors by integrating an effective feedback process into their systems. This is quite a remarkable success, as individuals cannot be expected to give or gains access to assets without trust.

Another important idea that drives the sharing economy is the maximum utilisation and distribution of idle but useable space, asset, and object or time. The evidence that seems to be emerging shows growth of this economy with many small, local imitative informal platforms emerging in the markets of developing countries where regulations are weak.

Sadly, the crucial issue of distributive justice has escaped the attention of the policy makers and the public at large. The dominant ideas of free-market economy, and neo-liberal framework have pushed down welfare, job protection and redistribution of wealth through effective taxation. Regulatory and social conflicts and disputes exist but those on the receiving end of the system, like the taxi drivers, find it difficult to raise their voices louder and to be heard in the corridors of power. The future outlook of the sharing economy looks promising in the sense of transitioning to mainstream economy, as online services and platforms become cheaper, easily available and affordable worldwide. The more it happens, the more we fear transfer of wealth from the periphery to the metropolitan economies and more concentration of wealth in the older economic classes and the new technological elite that seems to be emerging.

References

Abelson, R. (2016) "American Well will Allow Telemedicine Patients to Pick Their Doctor". *The New York Times*, May 16. Available at: https://www.nytimes.com/2016/05/17/business/american-well-will-allow-telemedicine-patients-to-pick-their-doctor.html?searchResultPosition=4. Accessed on June 24, 2020.

American Well (2019). AmericanWell.com. Available at: https://static.americanwell.com/app/uploads/2019/09/American-Well-Award-Frost-and-Sullivan.pdf. Accessed on June 24, 2020.

Bort, J. (2019). "Here's who is getting rich from Lyft's enormous IPO", *Business Insider*, March 29. Available at: https://www.businessinsider.com/lyft-ipo-paying-off-big-time-for-these-10-people-2019-2#over-100-others-13. Accessed on June 23, 2020.

Bourguignon, F. (2016). "Inequality and globalization: How the rich get richer as the poor catch Up", *Foreign Affairs*, January-February 2016.

Celment, J. (2020). Ebay's annual net revenue. *Statisa.com*. Available at: https://www.statista.com/statistics/507881/ebays-annual-net-revenue/. Accessed on June 23, 2020.

DAWN. (2020). "Project to enable patients to get online consultation". *Dawn*, March 27. Available at: https://www.dawn.com/news/1544018/project-to-enable-patients-to-get-online-consultation. Accessed on June 24, 2020.

Dillon, D.S. (2016) "Affordability, Efficiency and Convenience will Prevail", *The Strait Times*, August 15. Available at: https://www.nst.com.my/news/2016/08/165612/efficiency-reliability-affordability-and-convenience-will-prevail. Accessed on June 23, 2020.

Fitzsimmons, E. (2018). "A taxi driver took his own life. His family blames Uber's influence". *The New York Times*, May 1.

Forbes (2019). "Breaking down Lyft's Valuation: An Interactive Analysis", *Forbes*, May 7. Available at: https://www.forbes.com/sites/greatspeculations/2019/03/07/what-will-lyfts-ipo-valuation-be/#184b23db2040. Accessed on June 23, 2020.

Forbes. (2018). "Is $80 Billion Valuation Achievable for Didi Chuxing's IPO"? *Forbes*. Available at: https://www.forbes.com/sites/greatspeculations/2018/12/24/is-80-billion-valuation-achievable-for-didi-chuxings-ipo/#685f5f216211. Accessed on June 20, 2020.

Hannak, et al., A. (2017). "Bias in Online Freelance Market Places: Evidence from TaskRabbit and Fiverr". https://personalization.ccs.neu.edu/static/pdf/hannak-cscw17.pdf. Accessed on June 23, 2020.

Hardy, Q. (2015). "The Sharing Economy Visits the Backcountry", *The New York Times*, July 1. Available at: https://bits.blogs.nytimes.com/2015/07/01/the-sharing-economy-visits-the-backcountry/?searchResultPosition=1. Accessed on June 20, 2020.

Hayek, F.A. (1972). *The Constitution of Liberty*. Chicago: University of Chicago Press John

Iqbal, M (2020). "Uber Revenue and Usage Statistics", *Business of Apps*. Available at: https://www.businessofapps.com/data/uber-statistics/. Accessed on June 20, 2020.

Knvaul Sheikh, K. (2020). "Find a Vaccine. Next: Produce 300 Million Vials of It". *New York Times*, May 1. Available at: https://www.nytimes.com/2020/05/01/health/coronavirus-vaccine-supplies.html?searchResultPosition=2. Accessed on June 18, 2020.

Macrotrends.net. (2020) "Etsy Market Cap" *Macrotrends*. Available at: https://www.macrotrends.net/stocks/charts/ETSY/etsy/market-cap. Accessed on June 23, 2020.

Maheshwari, S. (2019). "Rent the Runaway Now Valued at $1 Billion With New Funding". *The New York Times*, March 21. Available at: https://www.nytimes.com/2019/03/21/business/rent-the-runway-unicorn.html. Accessed on June 23, 2020.

Mason, P. (2015). "Airbnb and Uber's sharing economy is one route to dotcommunism", *The Guardian*, January 21. https://www.theguardian.com/commentisfree/2015/jun/21/airbnb-uber-sharing-economy-dotcommunism-economy. Accessed on June 26, 2020.

Mays, J. (2018). "3 Years Ago, Uber Beat Back a Cap on Vehicles. What's Changed? A Lot". *The New York Times*, August 9. Available at: https://www.nytimes.com/2018/08/09/nyregion/uber-cap-nyc-decisionstrategy.html?searchResultPosition=5. Accessed on June 23, 2020.

McGinnis, D. (2018). "What is the Fourth Industrial Revolution" *Leadership Insights, Artificial Intelligence*, December 20. Available at: https://www.salesforce.com/blog/2018/12/what-is-the-fourth-industrial-revolution4IR.html#:~:text=The%20Fourth%20Industrial%20Revolution%20is,quantum%20computing%2C%20and%20other%20technologies. Accessed on July 1, 2020.

Miller, D. (2019). "The Sharing Economy and How it is Changing Industries", The Balance Small Business. https://www.thebalancesmb.com/the-sharing-economy-

and-how-it-changes-industries-4172234#:~:text=The%20sharing%20economy%20 is%20an,under%2Dutilized%20skill%20or%20asset. Accessed on June 19, 2020.

Nozick, R. (1973). "Distributive justice". *Philosophy & Public Affairs*, 3(1), pp. 45–126.

NYT (New York Times) (2014) "Conflicts in the Sharing Economy". *The New York Times*, November 4. Available at: https://www.nytimes.com/2014/11/05/opinion/ growing-pains-for-airbnb-and-others.html?searchResultPosition=3. Accessed June 25. 2020.

NYT (New York Times). (2014). "Brokers of the Sharing Economy". *The New York Times*, August 16. Available at: https://www.nytimes.com/2014/08/17/technology/ brokers-of-the-sharing-economy.html?searchResultPosition=4. Accessed on June 19, 2020.

Okenwa, I. (2020) "How Much is eBay Worth in 2019 and How Do They Make Their Money", *Just Riches*, April 6. Available at: https://justrichest.com/ebay-worth-in- 2019-how-they-make-money/. Accessed on June 23, 2020.

Pouspourika, K. (2019). "The Four Industrial Revolutions", *Institute of Entrepreneurial Development*. Available at: https://ied.eu/project-updates/the-4-industrial- revolutions/. Accessed on June 18, 2020.

PWC (Price Waterhouse Coopers). (2015). "Shared Economy: Intelligence Survey Series". *PWC*. Available at: https://www.pwc.fr/fr/assets/files/pdf/2015/05/pwc_ etude_sharing_economy.pdf. Accessed on June 23, 2020.

Rawls, *A Theory of Justice* (Princeton, New Jersey: Princeton University Press, 1971).

Roger, C.B. (2010) "Amartya Sen and the Idea of Justice" Open Democracy, October 5. Available at: https://www.opendemocracy.net/en/amartya-sen-and-idea-of-justice/. Accessed on June 30, 2020.

Schwab, K. (2016). "The Fourth Industrial Revolution: What it Means, How to Respond" *The World Economic Forum*, January 14, 2016. Available at: https://www.weforum. org/agenda/2016/01/the-fourth-industrial-revolution-what-it-means-and-how-to- respond/. Accessed on June 30, 2020.

Schwab, K. (2018). "The Fourth Industrial Revolution", *Encyclopedia Britannica*. Available at: https://www.britannica.com/topic/The-Fourth-Industrial-Revolution-2119734. Accessed on June 18, 2020.

Schwab, K (2018). The fourth industrial revolution. *Journal of International Affairs*, 72(1), pp. 13–16.

Singhal, S and Carlton, S. (2019). "The Era of Exponential Improvement in Healthcare"? *McKinsey & Company*, May 19. Available at: https://www.mckinsey.com/industries/ healthcare-systems-and-services/our-insights/the-era-of-exponential-improvement- in-healthcare#. Accessed on June 24, 2020.

TaskRabbit (2020) TaskRabbit.com. Available at: https://www.taskrabbit.com/. Accessed June 23, 2020.

UN (United Nations) (2020). "Rising inequality affecting two-third of the globe, but it is not inevitable". *UN News*. United Nations. Available at: https://news.un.org/en/ story/2020/01/1055681#:~:text=Inequality%20is%20growing%20for%20 more,by%20the%20UN%20on%20Tuesday. Accessed June 25, 2020.

Wootton, et al., R. (2012). "Long-running telemedicine networks delivering humanitar- ian services: Experience, performance and scientific output". *WHO Bulletin*. Geneva: WHO.

11

UNEVEN DEVELOPMENT, DISCRIMINATION IN HOUSING AND ORGANIZED RESISTANCE

Nima Hussein and Josh Hawley

A neighbourhood crisis

On a bright sunny Saturday afternoon in early June 2018, tenants from an Ottawa, Canada, neighbourhood known as Herongate packed a large meeting room at the local community centre. A tight-knit group of neighbours, mostly Somali and Arab, made up the majority of the gathering. They were there to discuss strategies to defend their homes and by extension the supportive enclave they had fostered over decades. Just over three weeks earlier, on May 7, their $10-billion corporate landlord notified them—all 105 households of them—of its plan to demolish their entire block of row townhouses.

The landlord, Toronto-based Timbercreek Asset Management, wrote in this notice: "the homes … are reaching the end of their building lifecycle … 25% of these homes are no longer viable" (Timbercreek Communities 2018). Tenants knew their homes were not condemned, even after the succession of corporate landlords had strategically neglected the property—cracked foundations were left unrepaired, original windows were never replaced and leaking roofs ignored, among other issues. Tenants knew their landlord was unwilling to spend money on repairs, but they also knew their three and four-bedroom brick townhouses, built in the late 1960s by Ottawa-based developer Minto in the *en vogue* French Eclectic Mansard style, were solid and sturdy. Tenants took care of them. Many families were raising their children there. Some older tenants had lived in their home for over 40 years, their children long-since moved out. What did Timbercreek mean by "no longer viable?"

At the meeting on June 2, while children drew portraits of their homes and wrote messages like "We are staying" and "We live here" on white Coroplast boards, the United Nations Special Rapporteur on Adequate Housing Leilani Farha, herself from Ottawa, stood at the front of the room and told tenants what

Timbercreek was doing was a gross violation of international human rights, all in pursuit of profits. She later told organizers of the meeting that when she stepped into the room, the demographic makeup of the tenants immediately struck her. People from the neighbourhood already were well aware. This was a neighbourhood of Black and brown people targeted for revitalization and redevelopment, to be replaced by higher-paying, primarily white tenants, the plan engineered by powerful developers and landlords, and with the full support of the Ottawa city government.

Over the course of the summer 2018, organizers with the Herongate Tenant Coalition—composed of tenants from Timbercreek's property and adjacent rentals—conducted an extensive census of that block. The results were staggering: 93% were People of Colour, and there were on average 5.4 people per house, meaning around 550 people, including over 200 children, were being pushed out. While the goal of the tenant committee was to prevent the demolition and for tenants to stay in their homes, all households were eventually forced out due to a variety of tactics employed by Timbercreek.

The extent to which Timbercreek, a financialized landlord with an extensive toolbox of tactics to increase their investment yields, maximizes profits and exploits marginalized tenant bases was soon revealed to tenants in Herongate. These tactics, and the driving forces behind them, are discussed in this chapter, as well as methods to impede and challenge them in the form of both direct action and policy changes, both of which reduce harm caused by capital accumulation. This chapter examines the uneven development spread across different geographies, in particular how exploitation and discrimination occur by consistently displacing the surplus labour supply and how immigration is the single most dominant fundamental (a market term denoting a factor that ensures profitability) in the rental housing industry in Canada.

Financialization of housing and global capital

Finance is an international system, what Greek economist Yanis Varoufakis likens to a global vacuum cleaner of capital (Varoufakis 2015). The impacts of globalization on cities manifests as a replication of the processes of economic production, which can be found in city after city around the world:

> "some cities—New York, Tokyo, London, Sao Paulo, Hong Kong, Toronto, Miami, Sydney, among others—have evolved into transnational market 'spaces.' As such cities have prospered, they have come to have more in common with one another than with regional centres in their own nation-states, many of which have declined in importance. ... [T]he impact of global processes radically transforms the social structure of cities themselves—altering the organization of labour, the distribution of earnings, the structure of consumption, all of which in turn create new patterns of urban social inequality" (Sassen 1994).

The industries that drive capital accumulation in one part of the world can be found (in some form or another) anywhere else on the planet. This section examines how the global mechanisms of the professionally managed real estate investment market (worth USD 9.6 trillion in 2019) (Teuben and Neshat 2020) and the broad pension funds sector (worth USD 32 trillion in 2019) (OECD 2020) are felt at a local district, neighbourhood and even building level.

Canada has the fifth largest pension fund industry in the world, valued at USD 1.5 trillion. Many of these funds, particularly those in the public sector, have ethical commitments around "socially responsible investing" which generally entail staying away from weapons, alcohol, and gambling, among other controversial industries. As institutional investors, large pension funds have found stable, and supposedly ethical, returns in the residential rental industry. International asset managers and real estate investment trusts (REITs) use multi-family assets (also known as apartment buildings) as financial vehicles to attract investors, such as teachers' and public service employees' pension funds, based on the relatively low capitalization rate (which is an indicator of risk) of rental housing.

The financialization of housing, then, "remakes homes into assets" (August 2020) by converting a basic human necessity—shelter—into a stable return on investment. These companies, such as Timbercreek (the largest landlord in Herongate), use the reliability of rent collection and the necessity of housing for survival as a key selling point: "When we are looking for ways to generate a predictable yield on an apartment building, where the rent comes from tenants who are paying for a roof over their heads, is far more predictable than owning a shopping mall where the tenants' ability to pay rent is driven by sales of a business" (Threndyle 2009). An executive with Swedish-based Akelius, which owns more than 47,000 apartment units across Sweden, Germany, Canada, England and France, told an industry magazine, "We have a responsibility to deliver basic human needs to our customers" (Rental Housing Business 2014). Individuals can also buy stocks of the landlord companies that are publicly-traded, effectively becoming partial owners of an apartment building. Western Canadian crown corporation the Alberta Investment Management Corporation (AIMCo) is a major investor in Toronto corporate landlord MetCap while the Teachers' Retirement Allowances Fund, the pension fund for teachers in the province of Manitoba, were partial owners of the Herongate property through one of Timbercreek's holding companies.

The financialization of housing can also be understood on a broader scale as the development of global inequality as a result of neoliberal, post-interventionist capitalism. The mainstream narrative surrounding postwar globalization is predominantly a discourse dictated by Western capitalist hegemony, in which the global South exists as a resource and labour rich economy for the plundering. Understood by many instead as an insidious by-product of Western imperialism and colonialism, this neo-colonialism is cancerous to the fabric of a righteous, healthy society with intentions for an equitable, sustainable future. Arguably, at the centre of many of these contemporary international struggles lies the unregulated free market, and the proliferation of capital over humanity, regardless of global frameworks attempting to mitigate uneven development.

Despite the severe detriment to indigenous populations, ancient natural infrastructure has suffered mass eradication and natural economies disrupted in the name of global economic progress and development. And so, one must ask: how many lives have been stolen for Western measures of progress? How many cultures have been destroyed? To this day, socio-political inequality proliferates under the veil of global economic prosperity, and the most vulnerable are exploited as underpaid labourers, risking death and dismemberment working in unregulated industries under shoddy and unsafe conditions for measly profits of which the bulk instead line the pockets of private industry owners, or even as victims of modern slavery, forced to work themselves to the bone with little to no financial compensation.

The neo-colonial influence on the larger global political economy, as well as the rampant economic exploitation of populations made vulnerable through both the deregulation of the free market and the complicity of the nation state, reveal an undeniable link between colonial powers basking in the newfound wealth of postwar globalization and former colonies suffering as a result of the economic disparity:

> In many developing countries, the post colonial period brought immense wealth to the elite and increased the impoverishment of the poor majority. … Meanwhile, traditional ways of life and subsistence are sacrificed to cash crops in order to service foreign debt. The processes of modernization and globalization in the world economy have had a profound impact on indigenous peoples and the small-scale subsistence farming which supported them. The forced shift from subsistence to cash-crop agriculture, the loss of common land and government policies which suppress farm income in favour of cheap food for booming urban populations have all helped to bankrupt millions of peasants and drive them from their land and sometimes into slavery (Bales 2000).

Tried and true imperialism and colonialism may purportedly be relics of the past, but their mutations have dictated and informed the politics and culture surrounding globalization and the broader international community. Frameworks attempting to mitigate the resulting strife and massive inequity have been brought forth by a multitude of intergovernmental organizations, but none so comprehensive as the United Nations' Millennium Development Goals and the subsequent Sustainable Development Goals.

The United Nations Sustainable Development Goals sprung into existence as a continuation of the Millennium Development Goals, an extensive set of targets addressing poverty, hunger, gender inequality, and other societal ills.

Each of the 191 member nation states consented to achieve the MDGs by 2015. When this was not the case, the United Nations General Assembly then set forth the 17 Sustainable Development Goals, which are described as inter-connected efforts in addressing global inequality by 2030. However, within a late stage capitalist society, establishing the unequivocal rights of all men is a daunting task, as the unregulated free market not only sanctions, but fun-damentally encourages profit to the highest degree. This particularly rings true for Goal 1 and Goal 11 of the SDGs doctrine as they relate to the current housing crisis, "ending poverty" and "making cities and human settlements inclusive, safe, resilient and sustainable" (United Nations 2020). As housing is increasingly financialized within global markets, putting an end to endemic disenfranchisement and adequately housing rapidly growing urban populations becomes increasingly arduous.

The fulfilment of the SDGs and any framework encouraging global human progress requires a distinct level of integrated processes, and despite the United Nations' uncontestable insistence of the nature of the goals as interlinked, there exists within the agenda no comprehensive strategy or approach to adequately employ these sustainable measures (Lempert 2017). With no real means to stimu-late international compliance or enforce the SDGs, it appears international actors are free to pay lip service and then continue their individualist agendas, once again prioritizing profit over people and furthering uneven development.

Immigration, labour and marginal areas

On the outskirts of Ottawa, a global superpower has set up shop. Amazon opened a one-million square foot warehouse in 2019 in Carlsbad Springs, a rural commu-nity with no public transit access. Every day, six hundred full-time workers head to the facility on Amazon Way. Amazon warehouse job advertisements hang in welfare offices across the city. It is these low-wage jobs, inaccessible by public transit, that ensure precarious and dangerous work will be available for new immigrants and tenants housed in large rental districts. Ottawa has announced the construction of a second Amazon warehouse in another far corner of the city.

The peri-urban area (throughout the literature is called the backcountry, hin-terland, diffuse urbanism, *Zwischenstadt* ("in-between city"), extended metro-politan areas) (Davis 2007) has consistently been the site of new industrialization. Two connected, but distinguishable, areas can be identified: the "far" hinterland and the "near" hinterland. The former is largely rural although it "also includes large urban zones of collapse, which exhibit almost identical characteristics." It is a site characterized equally by resource extraction and production as the preva-lence of the informal economy—black markets, illegal drugs and human traffick-ing. The near hinterland is largely suburban or peri-urban, from the sprawling suburbia's of North America to the *banlieues* of Europe to the slum cities of Africa and Latin America. It is here where most of the world's population lives (Neel 2018).

These outlier areas are crucial in understanding uneven development because this is where populations of wage earners and new immigrants predominantly live. Development occurs here not only to provide housing and accommodate the masses, but to exploit these populations using predatory practices of neglect and displacement. The perimeters of cities are where many migrants, both internal and external, first land and where they often take root and establish enclaves.

In Ottawa, Herongate was originally developed to house displaced working class people from the downtown area LeBreton Flats, near Parliament Hill, that was expropriated by the federal government in the name of urban renewal. The railway was moved from LeBreton Flats to the far south end of the city in 1964 (Picton 2009) and the working class population followed suit, as property on the margins is of low value to the ruling class. The federal land use planning agency commented on such industrial, working class areas:

> "The land value plan of Ottawa, Hull and vicinity is an excellent guide for the planner; its direct relation to the urgent problems to be solved is obvious: railroad situations, blighted areas, congested and unsanitary housing, are clearly incident to sections of the cities where land values are comparatively low. Improvement of such sections is therefore made possible, and, by fostering land revaluation, becomes a profitable operation" (Picton 2015).

As working class populations are repeatedly displaced over time for a variety of reasons, including the reduced profitability of maintaining older properties, the economics of new builds and the construction industry, the growth of cities, and the increasing need for capitalism's dispossessed surplus army of labour, landlords remain ready to seize this opportunity to cash in on crises. Rentals, especially apartments towers (referred to in the industry as multi-residential assets) due to their density and the economy of scale, are a main site of uneven development primarily driven by the business practices weaponized by landlords (August 2020).

Immigration, low-wage employment centres and the resource industry, in addition to lax tenant protection laws, are the forces that drive financial corporations into the rental housing market. In 1997, the Ontario legislature passed a law which abolished rent control on vacant units and introduced a loophole for landlords to increase rents on occupied units. This loophole took the form of the Above Guideline Increase (AGI), which allows landlords in Ontario to raise rents above the annual rent increase amount fixed to the government-determined Consumer Price Index. Landlords can submit applications for higher rent increases, which are almost always approved, to the Ontario Landlord and Tenant Board based on certain irregular expenses. These AGIs have spurred large tenant actions across Ontario in the form of rent strikes and public shaming campaigns against landlords. The first Real Estate Investment Trust (REITs are one of the main financial structures used in the financialization of multi-family rental housing), was formed at just this time (August 2020).

Global flows of immigration and patterns of uneven development are easily linked to Herongate in Ottawa. Timbercreek has been open about the role immigration and low housing supply play in ensuring high revenues (in the case of landlords, revenues are almost exclusively derived from monthly rents). In 2016, Timbercreek's co-founder Ugo Bizzarri said, "The demographics for rental is very strong. There's a lot of immigration. There's a lot of population demand. The seniors are selling their homes and they want to rent. From a real estate-fundamentals (perspective), rental is a very strong market" (Duggan 2016). Bizzarri is quoted as saying "that few apartments have been built over the past 40 years. At the same time, 'we are adding almost a million people every 10 years, and among newcomers to Toronto around 80% have a preference for renting'" (O'Dea and Schwartz-Driver 2014). By exposing Timbercreek on multiple fronts, through academia, the news media and on-the-ground organizing, it is now clear that "Timbercreek's profits are linked to displacement. ... 'the business plan ... is to roll the tenants'" (August 2020).

Timbercreek, the seventh largest landlord in Canada in 2017, is known to be one of the most flagrant actors when it comes to employing these tactics of financialized landlords. In Herongate, they have weaponized daily operations in the form of "active management." This involves evicting tenants—called "lease rollover"—to take advantage of vacancy decontrol and regular "contractual rent increases." It also means they will do cosmetic upgrades to their buildings rather than repairs to tenants' units: "Things like renovating an apartment building's lobby or adding high-end amenities can generate higher rents over the long term" (Timbercreek 2019).

The interplay between the cyclical displacement of working class populations, immigration, and the development of perimeter areas relative to city centres, is at the core of the majority of exploitation and uneven development. Commentary by bankers and real estate executives reveals how significant these factors are. As an industry magazine wrote in 2014, "For developers, immigration has been the secret saviour" (Goodman Report 2014). A senior executive at CBRE, a major real estate investment firm, told the same magazine, "intensification [is] finally happening in the traditionally marginal areas." The Deputy Chief Economist of CIBC World Markets Inc., the investment banking arm of the Canadian Imperial Bank of Commerce, the fifth biggest bank in Canada, said that given consistent immigration, "the propensity to rent will be higher than ever" (*ibid.*) In the US, immigration also constitutes a "critical contribution" to the rental and labour forces. Immigrants make up 20% of all renter households, and a third of trades workers (Joint Centre for Housing Studies of Harvard University 2019).

This interplay is also what creates the conditions for revolt and uprisings. The people on the margins are pushed until they cannot take it anymore. In China, electronics manufacturing giant Foxconn has built so-called "Foxconn cities" throughout the country, each employing tens of thousands to hundreds of thousands of employees. The largest of these "cities" is Longhua Town, on the outskirts of megacity Shenzhen, which consists of 15 factories. In September 2012, 2,000 workers at the "city" in Taiyuan rioted after experiencing abuse from police and unlivable conditions in the Foxconn dormitories, but the riot quickly "shifted the focus on to broader living conditions, particularly for migrant workers who live in thousands of factory dormitories around the country" (Martina 2012). The global pattern of further exploiting marginalized groups is a vicious cycle. In the case of Foxconn's operations in China, the surplus population of labour to feed its factories derives from internally displaced populations housed in massive complexes (Chuang 2017).

Unhoused

It comes as no surprise that societies across the globe are currently undergoing a widespread housing and homelessness crisis. Within Canadian cities alone, affordable and equitable housing is disappearing at a previously unseen rate, largely due to the unrestrained financialization of housing, and housing increasingly viewed as a commodity rather than a necessity. As a result, compounded by a lack of social systems acting as a safety net, the scale of homelessness has skyrocketed, forcing victims and their families into shelters or onto the streets.

With public expenditure for affordable housing thrust to the wayside, and developers pouring money into local city councils, pushing their own self-interest in lieu of public good, low income tenants must not only reckon with growing rents, but are routinely forced into precarious housing situations. These issues are further exacerbated by corporations strategically positioning ownership of housing as vehicles of profit, therein transforming housing into commercial streams of income for investors. Within the global free market, housing is viewed as a commodity, and so REITs, as well as asset management firms, are well within their rights to buy up vast swaths of land and pour millions of dollars into housing infrastructure, if only to expand their portfolio. In order to create a return on their investment, financialized landlords must raise rents at an astronomical rate, replacing disadvantaged, low income tenants with the wealthy, and more often than not, destroying working class communities of colour.

Within the city sanctioned urban revitalization of a Toronto neighbourhood, once a working class social housing project transformed into a mixed income neighbourhood for urban professionals, yet another microcosm of the broader housing crisis emerges. Regent Park's revitalization is symptomatic of the crisis; private developers buying up huge swaths of the housing market, allowing the dilapidation of low income units or buildings, and "revitalizing" the area by mowing down existing infrastructure populated by racialized, low income tenants, clearing ground and building luxury condos and housing units in the wake.

In 2015, Timbercreek Asset Management began the process of "demoviction" (demolition-evictions) in Herongate, an area predominantly populated by low income immigrants, refugees, and people of colour. The process involved the destruction of units and the eviction of dozens in the midst of a harsh winter. Despite a valiant effort in building working class power, and organizing amongst the community, an additional 500+ people were uprooted from their homes in 2018. These deracinated tenants received little to no assistance from the asset management firm, nor the city, instead left to fend for themselves within a housing market steeped in crisis. Instead, tenants turned to their neighbours for support and developed a strategy to fight the demovictions. While Timbercreek was intensifying the daily struggles of tenants and the housing crisis deepened in Ottawa, the company received over CAD 1.7 million from the City of Ottawa in the form of municipally funded grants from 2012 to 2019 (City of Ottawa n.d.).

Consequently, homelessness is a rapidly growing crisis, a societal ill unlimited to a single race, gender or creed, but endemic in some populations more than others. Within Canadian cities, groups most affected by homelessness exist as customarily the most vulnerable, including Indigenous women, who are further victimized by existing structures of colonialism, white supremacy, and gendered violence. Black and brown immigrants also contend not only with socio-political inequality, but struggle with economic disparities compared to their white counterparts.

Moreover, groups experiencing homelessness report suffering adverse health effects as a result, including addiction, injury, and sexual assault. Of homeless adults in Toronto, 40% have reportedly been the victim of a physical assault, and a further 21% of women experiencing homelessness have reported rape (Frankish, Hwang and Quantz 2005). Poverty has a direct association with increased substance use, and with a lack of local policy or social services providing safe harm reduction sites or encouraging the usage of clean needles and other safety apparatuses, people experiencing homelessness are placed at even deeper risk. The mortality rate of homeless groups in comparison to folks with stable housing is also greatly concerning. The rate of death for homeless girls in Montreal alone is a shocking 31% higher than the general population. In Ottawa, families have seen the highest rate of homelessness, an increase of 55% since 2014, well over the other demographic groups of single men, single women and single youth (City of Ottawa 2019). During this same period, around 230 three and four-bedroom townhouses were demolished in Herongate by financialized landlord Timbercreek Asset Management with the full support from all levels of government. While developers and politicians shout from the rooftops that more construction of housing is needed, the fact is that the housing and homelessness crisis is entirely manufactured and a product of the capitalist system. In Ottawa alone, an estimated 20,000 properties sit vacant (Hopulele 2019).

Faced with huge power imbalances and intensified stresses on exploited peoples, how are these practices resisted? In the following section, a clear program of community organizing and policy reform is put forward.

Building power towards liberation

At the policy level, there is a strong link between rent decontrol, reduced tenants' protections laws, and the proliferation of financialized landlords and the
predatory practices they employ to maximize return on their investments. In
the province of Ontario, the so-called Tenant Protection Act of 1997 introduced
vacancy decontrol and a rent increase loophole, setting the stage for financialized
landlords to proliferate. In 2020, Ontario introduced the Protecting Tenants and
Strengthening Community Housing Act, which "aims to speedup evictions by
limiting tenants' legal defences and in some cases removing the requirement to
hold eviction hearings" (Parkdale Organize 2020).

The financialization of housing as a field of study—primarily the domain of
geographers, economists and sociologists—has provided analyses that have been
uniquely useful to tenants who are organizing against evictions and rent increases
imposed by these corporate landlords. By understanding the tactics their landlords use and that the neglect, rent increases, harassment and displacement so
many tenants experience on a daily basis are the foundation of their landlords'
business model, they have been able to unite and build collective power.

Working class power, in its truest form, need not be derived from liberation
theory and its various off-shoots. At its core, this undisputed form of resistance is
a power harnessed from the ground up, often by populations with limited access
to academia. Instead, these working class communities create their own grassroots movements with the intent to liberate themselves from the crushing tyranny they contend with in their daily routine. True revolution is not born within
the halls of academics, nor the offices of the political elite, never! Revolution
occurs when the most displaced and the most exploited harness the indomitable
spirit of the oppressed and create change.

Some movements have been successful while explicitly contending with the
manufactured limitations of a racial capitalist society, and undertake unto themselves a radical reimagining of a just, equitable future. These movements must
centre the health and care of their interlinked communities, and dig into the definition of creating direct engagement through grassroots organizing. Immersed
in the Black resistance movement against racial discrimination and capitalism,
the 1970's Black Panther Party developed one of its most influential initiatives,
the Free Breakfast Program for children (Lateef and Androff 2017). Resistance at
its core means taking up the cause, doing the work, and creating a space unbound
by colonial, carceral, and anti-black notions of policing and surveillance. These
revolutionary frameworks are inherently rooted in the stripping of fear and the
transmission of devotion, where the most oppressed among us is able to easily
access not only the resources, but the social services she may require, where a
commitment is made and honoured in recognizing and condemning interlock-
ing dominations.

Intersectional liberation begins within a grassroots framework, and engages
in mutual aid approaches that directly benefit communities made vulnerable

through youth programs, capacity building, food banks, and similar initiatives. In developing understandings of transformative justice and critically analysing state sponsored biases, liberation is not only within sight, but near. Moreover, community justice is strengthened through amplifying the voices of the most disenfranchised, affecting civil disobedience, and maintaining solidarity with other oppressed peoples. It continues when there is a recognition that yes, the work starts with the people, but it starts primarily within the people, and from there, radical politics are worthless if they are not centred around the well-being of communities, mutual aid, and justice.

A plan of action to counter the mass demovictions that have been occurring in Herongate can be found by looking to struggles in other neighbourhoods, in both the past and present. In 1968, tenants in the Milton-Parc neighbourhood of Montreal, directly to the east of McGill University, formed an independent organization, the Milton-Parc Citizens' Committee, to refuse a large developer's plan to acquire and demolish an entire six-block area consisting mainly of historic triplexes home to working class and new immigrant tenants, to build large-scale, modernist, luxury apartments. Tenants occupied the developer's office, held demonstrations, stopped traffic, and squatted boarded-up homes. The fight lasted a decade, and while one section was demolished and phase one of this massive project was built, tenants were eventually successful, took ownership and converted their homes into the largest conglomeration of neighbourhood housing co-operatives in the country (Hawley and Roussopoulos 2019).

After years of victories both big and small in the Parkdale neighbourhood of Toronto, including two successful rent strikes, preventing evictions and supporting unionizing workers at the Ontario Food Terminal, Parkdale Organize, the working class group behind these actions, proposes four principles for community organizing, outlined in a report published by a local community legal clinic: direct engagement, district-based scale, struggles for daily life, and independent organizing (August and Webber 2019).

Direct engagement involves bringing demands directly to individuals and institutions who are responsible for the decisions and actions that further harm or exploit working class people. Keeping the focus at a local district, neighbourhood or apartment building level capitalizes on the energy and connections that neighbours have amongst each other and avoids broader efforts that "can dilute attention to Parkdale's struggles, and siphon off the direct energy needed to win achievable, neighbourhood-level goals." Struggles for daily life captures the need for simply talking to people who are struggling in society, call a meeting and develop a strategy with clear demands. Finally, remaining separate from state actors and social agencies allows members to remain focused on their priorities and engage in extra-legal approaches that contribute to success: "independence in decision-making, agenda setting and all aspects of organizing are necessary for this work to succeed."

Another working class group, called AngryWorkers, that bases themselves on the margins of west London, UK, in order to work in the low-wage and manual labour sectors, takes up a similar framework for addressing exploitation to that described in the Toronto report. While their principles align, they articulate their strategy as a rootedness in all aspects of working class life: "A revolutionary organisation should exist and act within the class, not in its place, or as outsiders. The program doesn't exist on paper. ... Only by basing our politics on direct experiences like this, where we are putting down roots in working class areas, rather than just knocking on their door when election time swings around, can we build a real, grassroots counter-power—one that actually involves working class people!" (AngryWorkers 2020).

The Toronto report also provides some clarity of how success is evaluated. Winning demands is obviously a key indicator. If tenants or workers have prevented an eviction or forced a landlord to drop a rent increase, their efforts have been successful. Laying the foundations of an organized body of exploited people was another indicator, as was building a sense of community and developing support from outside groups including the media, academics and lawyers. Aside from winning demands, empowerment is perhaps most significant. Razia and Habon, tenants interviewed in Herongate for this report, "with support from a translator, explained that 'we felt helpless because they are a company and they have more power, it was like we couldn't fight back.' After connecting with HTC [Herongate Tenant Coalition] and learning their rights, they felt better: 'we were scared, but not as much anymore'" (August and Webber 2019). The work undertaken by tenants in Herongate will hopefully have lasting results: "The Herongate Tenant Coalition articulates a different reality to the narrative of Canadian benevolence, embodying class solidarity, denouncing structural racism, and recruiting allies to donate to their legal case against Timbercreek Management" (Gomá 2020).

Even while using these indicators, whether success has been achieved by organized working class people can be a matter of perspective. Both authors of this chapter are founding members of the Herongate Tenant Coalition (HTC) and have spent the majority of their lives as renters in the neighbourhood. While tenants in Herongate were not able to stop Timbercreek Asset Management's demolition of townhouses, their on-the-ground organizing work has seen some positive results. HTC's mobilization and exposure through adept news and social media strategies have been effective at reframing the perception of the neighbourhood from a negative stigmatization to one of resistance and radical politics (Xia 2020).

Tenants have built several legal cases stemming from the mass demovictions of 2018. One is a large human rights case involving close to 40 tenants (Trinh 2019). After HTC spent almost a year collecting work orders, documenting neglect, interviewing tenants, fundraising, filing Freedom of Information requests, and sifting through corporate documents, among other tasks, both Timbercreek and the City of Ottawa are being brought before the Human Rights Tribunal of Ontario for a case that details a pattern of systemic neglect targeting a mostly

Somali and Arab tenant population. With the focus on the collective rights of a neighbourhood, this case challenges the normative liberal democratic notion of individual rights and aligns more with the framework of Indigenous rights (Kulchyski 2013). It also relies heavily on Timbercreek's own statements regarding their predatory business practices, including those described earlier in this chapter. Civil cases have also been filed against Timbercreek, for a campaign of defamation against organizers (Crosby 2019) and for a situation stemming from an executive lying under oath to a secretive City of Ottawa committee to avoid a costly repair to a townhouse (Rockwell 2018). The outcomes for all these cases have yet to be determined (they are stuck in a backed-up and bureaucratic legal system), but the level of organization and coordination, particularly among a network of organizers, researchers and lawyers, to build each of these cases has been impressive. All of these cases, in addition to the mobilization of tenants in direct confrontation with their landlords, could have a resounding impact on the future of financialized landlords in the province of Ontario, and potentially on a broader level.

Conclusion

The flows of finance are global, and the impacts are felt locally. While many of the tools in pursuit of capital accumulation are employed globally, they are honed to the specific conditions and legal contexts in which they operate. A case in point is the proliferation of financialized landlords across Canada. The number of real estate investment trusts (REITs) in each province are directly tied to provincial rent control and tenant protection laws. Ontario, for example, has weak rent control and fewer apartments compared with its neighbour, Quebec, but three times as many REIT-owned suites (August 2020).

Politicians attempt to pass off responsibility for the damage wreaked in the wake of global finance. This was the experience in Herongate. On June 28, 2018, tenants facing the destruction of their neighbourhood brought their demands straight to Ottawa mayor Jim Watson. His boardroom table was surrounded by staff, city lawyers, housing bureaucrats, and the local councillor. Six tenants and members of the Herongate Tenant Coalition were present. The city refused to acknowledge any responsibility for the displacement of hundreds of racialized tenants, deferring instead to the authority of the province for jurisdiction over evictions. The reality, as the working class is fully aware, is that it is our neighbours who are torn away. The housing and homelessness crisis is felt at the city level. Hotels in Ottawa are converted into emergency family shelters run by the city, where children return from school to a small room shared by their family and the parents do not have a kitchen to cook in.

Exploited and oppressed peoples only have one place to turn, and that is themselves. By collectivizing the common struggles amongst the working class, the power imbalance foundational to class society can be tilted in our favour. Whether this looks like direct confrontation, public shaming campaigns, or legal campaigns, the activities of organized peoples in the interest of their own liberation exposes the market system for what it truly is.

References

AngryWorkers. (2020). Introduction. In: *Class Power on Zero Hours* [online]. PM Press. Available from: https://classpower.net/intro/ [accessed 30 June 2020].

August, M. (2020). The financialization of Canadian multi-family rental housing: From trailer to tower. *Journal of Urban Affairs*, pp.1–23.

August, M. and Webber, C. (2019). *Demanding the Right to the City and the Right to Housing (R2C/R2H): Best Practices for Supporting Community Organizing* [online]. Toronto: Parkdale Community Legal Services. Available from: https://www.parkdalelegal. org/news/demanding-the-right-to-the-city/ [accessed 30 June 2020].

Bales, K. (2000). Expendable people: Slavery in the age of globalization. *Journal of International Affairs*, 53(2), pp. 461–484.

Chuang. (2017). Working for Amazon in China, where the global giant is a dwarf. *Chuang* [online], 4 April 2017. Available from: http://chuangcn.org/2017/04/working-for-amazon-in-china-where-the-global-giant-is-a-dwarf/ [accessed 30 June 2020].

City of Ottawa. (n.d.). *100% Municipally funded grants and contributions financial information and notice* [online]. Available from: https://ottawa.ca/en/city-hall/funding/community-and-social-support-funding/100-municipally-funded-grants-and-contributions-financial-information-and-notice#2017-100-municipally-funded-grants-and-contributions-financial-information-and-notice-community-and-social-support-funding [accessed 30 June 2020].

City of Ottawa. (2019). *10-Year Housing and Homelessness Plan, 2018 Progress Report* [online]. Ottawa, ON: City of Ottawa. Available from: https://documents.ottawa.ca/sites/documents/files/Homelessness-Report-ENG_2018.pdf [accessed 30 June 2020].

Crosby, A. (2019). Financialized gentrification, demoviction, and landlord tactics to demobilize tenant organizing. *Geoforum*, 108, pp.184–193.

Davis, M. (2007). *Planet of Slums*. London: Verso.

Duggan, E. (2016). Apartment building boom takes hold in Vancouver. *Renx.ca* [online], 6 April 2016. Available from: https://renx.ca/apartment-panel-vancouver-forum/ [accessed 30 June 2020].

Frankish, C. J., Hwang, S. W. and Quantz, D. (2005). Homelessness and health in Canada: Research lessons and priorities. *Canadian Journal of Public Health*, 96(S2), pp. S23–S29.

Gomá, M. (2020). Challenging the narrative of Canadian multicultural benevolence: A feminist anti-racist critique. *OMNES: The Journal of Multicultural Society*, 10(1), pp. 81–113.

Goodman Report. (2014). *Goodman Report: The Newsletter for Apartment Owners since 1983* [online], November, p.3. Available from: https://goodmanreport.com/app/uploads/2019/05/2014-Canadian-Apartment-Investment-Conference-no-crop.pdf [accessed 30 June 2020].

Hawley, J. and Roussopoulos, D., ed. (2019). *Villages in Cities: Community Land Ownership, Cooperative Housing, and the Milton Parc Story*. Montreal: Black Rose Books.

Hopulele, A. (2019). *"Ghost" Homes Across Canada: A Decade of Change in 150 Cities* [online]. Point2 Homes. Available from: https://www.point2homes.com/news/

canada-real-estate/ghost-homes-across-canada-decade-change-150-cities.html [accessed 30 June 2020].

Joint Centre for Housing Studies of Harvard University. (2019). *The State of the Nation's Housing 2019* [online]. Cambridge, MA: Harvard University. Available from: https://www.jchs.harvard.edu/state-nations-housing-2019 [accessed 30 June 2020].

Kulchyski, P. K. (2013). *Aboriginal Rights Are Not Human Rights: in Defence of Indigenous Struggles*. Winnipeg: ARP Books.

Lateef, H. and Androff, D. (2017). "Children Can't learn on an empty stomach": The Black panther Party's Free breakfast program. *Journal of Sociology & Social Welfare*, 44(4), pp. 3–17.

Lempert, D. (2017). Testing the global Community's sustainable development goals (SDGs) against professional standards and International law. *Consilience*, (18).

Martina, M. (2012). China's dorm room discontent emerges as new labour flashpoint. *Reuters* [online], 27 September 2012. Available from: https://www.reuters.com/article/us-china-foxconn/chinas-dorm-room-discontent-emerges-as-new-labor-flashpoint-idUSBRE88Q1QS20120927 [accessed 30 June 2020].

Neel, P. A. (2018). *Hinterland: America's New Landscape of Class and Conflict*. London: Reaktion Books. pp.17-18.

O'Dea, C and Schwartz-Driver, S. (2014). City Focus, Toronto: Cranes on the skyline. *IPE Real Assets* [online], March/April. Available from: https://realassets.ipe.com/real-estate/city-focus-toronto-cranes-on-the-skyline/10001301.article [accessed 30 June 2020].

OECD. (2020). *Pension Funds in Figures* [online]. Available from: http://www.oecd.org/daf/fin/private-pensions/Pension-Funds-in-Figures-2020.pdf [accessed: 30 June 2020].

Parkdale Organize. (2020). *Stop the Eviction Bill* [online], 26 May. Available from: http://parkdaleorganize.ca/2020/05/26/stop-the-eviction-bill/ [accessed 30 June 2020].

Picton, R. (2009). *"A capital experience": National urban renewal, neoliberalism, and urban governance on LeBreton Flats in Ottawa, Ontario, Canada*. PhD Thesis, University of Toronto.

Picton, R. (2015). Rubble and ruin: Walter benjamin, post-war urban renewal and the residue of everyday life on LeBreton flats, Ottawa, Canada (1944–1970). *Urban History*, 42(1), pp. 130–156.

Rental Housing Business. (2014). European Style Meets Canadian Living. *Rental Housing Business* [online], April, pp.10–20. Availabe from: https://issuu.com/profilemag/docs/akelius [accessed 30 June 2020].

Rockwell, N. (2018). Patterns of Neglect: Heron Gate residents suffer from a legacy of poor maintenance and landlords' use of a municipal committee to overturn bylaw orders. *The Leveller* [online], 15 November 2018. Available from: https://leveller.ca/2018/11/patterns-of-neglect/ [accessed 30 June 2020].

Sassen, S. (1994). *Cities in a World Economy*. Thousand Oaks, CA: Pine Forge Press. p.xiv.

Teuben, B and Neshat, R. (2020). *Real Estate Market Size 2019* [online]. New York: MSCI. Available from: https://www.msci.com/documents/10199/035f2439-e28e-09c8-2a78-4c096e92e622 [accessed: 30 June 2020].

Threndyle, R. (2009). Timbercreek Finds Value in Overlooked Properties. *Canadian Apartment Magazine* [online], February, pp.22–34. Available from: https://issuu.com/rick11/docs/cam_feb09_issue [accessed 30 June 2020].

Timbercreek Communities. (2018). *Re: Important Notice About Your Tenancy*. Document distributed to tenants at a meeting held by Timbercreek on 7 May 2018.

Timbercreek. (2019). *REITs as a proxy for direct real estate investing* [online]. Available from: https://timbercreek.s3.amazonaws.com/docs/default-source/white-papers/reits-as-a-proxy-for-direct-real-estate-investing.pdf?sfvrsn=864b552f_2 [accessed on 30 June 2020].

Trinh, J. (2019). "I'm sad and I'm struggling, but I'm going to fight" — Inside the Human Rights Tragedy at Heron Gate. *Ottawa Magazine* [online], 4 April 2019. Available from: https://ottawamagazine.com/people-and-places/fighting-for-rights-at-heron-gate/ [accessed 30 June 2020].

United Nations. (2020). *About the Sustainable Development Goals* [online]. Available from: https://www.un.org/sustainabledevelopment/sustainable-development-goals/ [accessed 30 June 2020].

Varoufakis, Y. (2015). *The Global Minotaur: America, Europe and the Future of the Global Economy.* London: Zed Books.

Xia, L. (2020). *Immigrants' Sense of Belonging in Diverse Neighbourhoods and Everyday Spaces.* MA thesis, University of Ottawa.

12

THE FEAR FOR ANOTHER REVOLUTION/COLONIALISM

The evolution of the monroe doctrine as an instrument of racist domination and hegemony in the caribbean

Juan Velásquez Atehortúa

The Covid-19 pandemic outbreak seems to have changed the history of humanity in a before and after it. Upon its beginning, environmentalists were happy of the return of wild animals to urban centers and the halt of global emissions. But when the number of dies and infected grew, it was evident that black and coloured populations in the North Atlantic powers were the most affected. The situation framed a powerful return of antiracist and anti-colonial mobilizations headed by the #BlackLivesMatter movement. At the same time the North Atlantic empires used their military power to repress their own populations and to capture medical devices on the seas. Moreover, they intensified their economic sanctions with a naval blockade against the Caribbean nations that had risen against their hegemonies. As during the colonial piracy of 1700s, the North Atlantic powers coordinated how to confiscate gold reserves, bank accounts and subsidiary companies with operations in the US and Europe. In open violation of the charter of the United Nations, its financial institutions followed the Monroe Doctrine, a consensus between the US and its North Atlantic allies to intervene in the countries of the Americas. This time the doctrine was used to cut off access to credit to Venezuela and Cuba for the acquisition of medicines and hospital equipment. For their part, these Caribbean nations controlled the pandemic focusing on the Cuban health prevention model developed since the 1960s. Cuba went even further and lent its medical personnel to some 60 countries in the global south and in Europe (Fitz, 2020), turning medical care into an instrument of solidarity with whom Frantz Fanon (1967) calls "the wretched of the earth". But as if very little had changed in 200 years of history, the response from the liberal regimes to Cuba and Venezuela in Europe and the Americas continued to be indifference, harassment and racism.

This chapter addresses the rise and fall of the Monroe doctrine as an instrument of North Atlantic colonial dominance. As will be exemplified below, through the years the Monroe doctrine has developed into an international consensus among North Atlantic white colonial elites to freeze black and brown populations in Caribbean and Latin American societies in a perpetual state of exploitation and submission to their interests. However, at the time of the Covid-19 pandemic, the already active support of Russia and China to Venezuela (Pantoulas and McCoy, 2019) and to Cuba may have disrupted this hegemony to the grade of contesting the North Atlantic powers military and economic dominance. Is this the end of the North Atlantic dominance in the Caribbean, or is this the start of yet another era of much more enlarged bourgeois or liberal imperial dominance?

To answer this question, the chapter recounts some historical events around 1803, 1903, and 2003 until today, to stress the domination of North Atlantic racism to ensure the hegemony of the Euro-American civilization project. In the next section the text combines these temporalities around the relationship between the domination from Foucault and hegemony from Gramsci to show how racism serves to perpetuate its coloniality around the "Fear of turning into another" Haiti, Cuba, or Venezuela.

Domination and hegemony

The idea of domination and sovereign power as developed by Michel Foucault (2003) sustains that the theory of sovereignty shows how power can be constituted on the basis of a form of basic legitimacy that is more fundamental than all laws (Foucault, 2003, pp. 16-17). He observed that the social relations of Western societies follow two schemas that complement each other. One he calls the *war -repression schema* where opposition to the dominant power does not go beyond the legitimate or illegitimate. But beyond struggle or submission. Here different enemies fight each other in open wars. The one who has control over the resources of the state tries to dominate its enemy with open repression. In the *contract-of-oppression-schema*, the model freezes the balance of power from the last victory battle by moving this military dominance in the legislation.

In this text I want to use Foucault's two schemas to frame the colonization of the Caribbean as a project of dominance. With the imposition of slavery, for organizing the extraction of profit to establish global capitalism, the North Atlantic colonial powers imposed a contract of oppression model over the black and brown populations of the Americas and Africa. I coincide with Foucault in the sense that since the imposition of slavery in the Caribbean region theories of law have played a fundamental role in legitimizing political power by highlighting sovereignty's way of forcing the legal duty to be obedient. In the cases at issue, the North Atlantic empires use the Monroe doctrine in its foreign policy to claim sovereignty over the labor, territories and resources of Afro-Latin and indigenous populations.

However, my perception is that applying Foucault's models of dominance claims further sophistication, for example by looking at it in relation to hegemony. According to Joseph Femia (1975, p. 29) Antonio Gramsci's perception of hegemony refers to "a situation wherein a social group or class is ideologically dominant." In this line, writes Femia, the supremacy of a social group manifests "through the coercive organs of the state, and 'intellectual and moral leadership', which is objectified in and exercised through the institutions of civil society, the ensemble of educational, religious, and associational institutions" (Femia, 1975, p. 30). Here, following Gwyin Williams' definition, also based on Gramsci's writings, Femia states that hegemony consists in

> "an order in which a certain way of life and thought is dominant, in which one concept of reality is diffused throughout society, in all its institutional and private manifestations, informing with its spirit all tastes, morality, customs, religious and political principles, and all social relations, particularly in their intellectual and moral connotations" (Williams 1960, p 587, quoted in Femia, 1975, p. 31).

From this quotation Femia arrives to the insight that Hegemony is therefore "the predominance obtained by consent rather than force of one class or group over other classes." (Femia, 1975, p. 31). My perception is that Gramsci's notion of hegemony cannot be disconnected from the models of domination advanced by Foucault. What this article aims is to use domination and hegemony to elucidate why anti-imperialism, although a necessary struggle, no necessarily is decolonial and can even contribute to secure the hegemony of the bourgeoise class when facing alternative futures to its own vision of society. In order to make the anti-imperialist effort congruent with a decolonial pursue we need then to better understand the relationship between domination and hegemony among the social constellations assuming the roles of being the subjects of history. In the next sections the chapter illustrates examples in this regard aiming to confirm this perception.

Dominance and hegemony in 'the fear for another Haiti'

In 1800, most of the sap with which, then the island of Santo Domingo, sweetened the economy of Europe came out from the sugar cane and coffee fields of Haiti. By the time of the outbreak of the French Revolution in 1789, Haiti concentrated 40 percent of France's international trade, exporting half of the world's coffee and quantities of cane sugar similar to the production of Jamaica, Cuba and Brazil together (Shah, 2009, p. 24). But as the value of their production grew, so did the resistance against slave labour. This resistance went in two directions according to de Melo Rosa and Pongnon (2013). One direction was defended by Toussaint L'Ouverture, who, inspired by the ideas of the French Revolution, sought to establish an alliance with it to add the island to the empire of the emerging French bourgeoisie. In light of the new liberal hegemony the main goal of L'Ouverture was to elevate the status of blacks, from slaves to the category of free citizens. The other direction, defended by Jean Jackes Dessalines, sought to separate itself entirely from white domination to establish a Black Empire projected against slavery and the racialization of labor relations. Due to his pro-French positions, L'Ouverture became the henchman of Napoleon Bonaparte, who re established slavery in the French colonies, assigned L'Ouverture the post of Governor General of Saint Domingo from 1795 to 1802. To maintain economic autonomy L'Ouverture continued sugar production, calling on free men to return to sugar cane fields under the same conditions of exploitation and giving whites the privilege of continuing to organize and head the production (Gorender, 2004, p. 299). The black population took away its support from L'Ouverture, opening the way for the abolitionist side of Dessalines to assume power in an insurrection in 1802.

Under this new situation, Dessalines prepared the declaration of independence from France in 1804 with which he began the construction of the empire of the emancipated black diasporas from the Caribbean nations. In that same year, he assigned land to farm and US $53 as a reward to anyone who escaped from slavery in the Caribbean islands and nations around to settle in Haiti. The Dessalines imperial project initially repatriated more than 13,000 fugitives who escaped from slavery (de Melo Rosa and Pongnon, 2013, p. 463). But the Black Empire had its first backslash with the assassination of Dessalines in 1806 at Pont Rouge, north of Port-ou-Prince. Dessalines's death gave rise to a faction in 1807 among the north, led by Henry Christophe and subscribed to the pro-French views of L'ouverture; and the south, led by Alexandre Pétion, who continued Dessalines' vision. In 1811, the northern part, under the command of Christophe, became a kingdom. While the southern part, under the leadership of Pétion, continued its path as a Black Republic (Ferrer, 2012). Ferrer writes how both parties developed economically:

> [While the north] retained large-scale landholdings, which were managed by military officers and produced sugar with the "attached" labor of former slaves. In the south, which was organized as a republic, the government had divided the large estates and carried out an agrarian reform, dismantling the old plantation system and distributing almost 100,000 hectares of land, mostly in modest plots of 25 to 45 hectares (Ferrer, 2012, p. 44).

However, both factions agreed in between the military, distributing among themselves control of the confiscated lands from the white elites, to maintain military loyalty to their general commanders. The black proletarians, being free in theory, thus saw their dreams vanished regarding acquiring land in which to cultivate and going towards a society free from the racialization of labor relations. In the light of Foucault's model, it could be said that the Haitian revolution marked a break with colonial domination and called into question white hegemony in the conduct of the construction of society. But in the light of an analysis of the hegemony of the capitalist system, according to Gramsci's thesis, the revolution was unable to break neither with the hierarchical organization of labor, nor with the subordination of Haiti in the global market as a producer of sugar and coffee. For these purposes, Haiti needed to surround itself with a community of nations to continue exploring alternatives to survive the North Atlantic counterattack.

Despite these hard setbacks, the colonial rule did everything possible to avoid being infected by the radicalism of the black Republic of Haiti (Romero Jaramillo, 2003, p. 22). Then "the fear for another Haiti" went around the independency leaders as well. In 1815, Haitian President Alexandre Pétion granted asylum to Simón Bolívar, who later received thousands of muskets, ammunition, food, and a printing press on the condition to advocate the abolition of slavery in the liberation struggle that would presumably start the post-imperial era. In that same year the Spanish crown promised the slaves freedom in exchange for defending the colonial monarchical power. However, the independence barons rushed to publish their offer of abolition to be one step ahead to attract the slaves to their independence army (Romero Jaramillo, 2003, p. 22). Bolívar showed a great commitment to make the black population acquire their freedom by joining the liberating forces. But because his generals refused to have blacks in their military ranks, he wondered:

> "Is it not more appropriate that they fight for their rights on the battlefield? And that their dangerous number is reduced by such a powerful and legitimate means? We in Venezuela have seen the free population die, while the captive exists in great number" (Chambers, 2006, p. 30).

From their oligarchic and bourgeois positions, and to avoid a continental revolt of slaves, the generals of the future Creole states promised to abolish slavery offering legalistic promises that they never kept. Bolivar's army managed to militarily defeat the imperial forces between 1819 and 1825 in all the Andean countries from Venezuela to Bolivia. With the fear that the experience of Haiti would be repeated in the new free republics, the Colombian Congress imposed in 1821 the concept of the free womb. This gave freedom to the unborn, but granted slaves conditional freedom only if they rendered their services to their masters or to the state, through military service. Slavery was only formally abolished in 1852 in Colombia and 1854 in Venezuela (Monguey, 2012, p. 54).

At the time of this controversies the US president James Monroe proclaimed its doctrine, in 1823, aiming to prevent the reconquering of Spanish colonies in the Americas with the support of the Saint Alliance, conformed by Austria, Russia and Prussia (Gramsci, 1981, pp. 217–218). Here is interesting to stress that despite the confluence of Monroe and Bolivar in its anti-imperial efforts, the ghost of "*The fear for another Haiti*" established a parallel hegemony to exclude Haiti from the first Pan American Congress of Independent States in Panama in 1826. This was just when the north Atlantic Empires imposed an economic and commercial blockade that demanded compensation for slave owners, for the 'loss' of their assets.

The Bonapartist state estimated this debt at 150 million francs to be transferred from state to state due to the loss of "servile property" in addition to the territories and instruments of production that, according to the commission sent to the island in 1825, amounted to one billion francs (Beauvois, 2010, p. 615). These measures were monitored by Colombia, which was then actively participating in the consolidation of an insular federation in alliance with Mexico with the ambition to establish a Cuban republic and a Puerto Rican republic. By that time the southern part of the Haitian Republic had annexed Santo Domingo, the oriental part of the island. Haiti was then included in Bolívar's plan to form a league of nations to repel the imperial attacks. For the Congress of Panama on August 11, 1826, Bolivar had three goals: to protect the American coasts, to launch an expedition against Havana and Puerto Rico, and "to march to Spain if the Spaniards did not want peace by then" (de la Reza, 2015, p. 75). The American union, headed by slave owners, was opposed to recognizing Haiti as an independent state. Nevertheless, it became at the end invited to this conference, while Haiti, which had contributed directly to the Bolivarian emancipation project, become excluded.

De la Reza has outlined five reasons for excluding Haiti from the conference. First, that this was for exclusively Spanish-speaking nations. But this is a position essentially contradictory since they also invited the American Union. Second, the American Union proposal, opposed by Mexico, that Colombia should desist from the liberation of Cuba and Puerto Rico, on the condition that Spain give a perpetual truce to the rest of the Americas. In this direction Bolívar apparently wanted to avoid a direct conflict with the American union, since its deputies, as slaveholders, supported the compensation demands imposed by the French empire against Haiti. And fourthly and maybe most important was the fear of more slave revolts in the Caribbean (de la Reza, 2015, p. 76). At the end, none of the delegates and the nations represented at the Panama conference recognized Haiti as a free nation. In other words, although Bolivar and his generals were anti-imperialist when it comes to Spain, their positions for excluding Haiti from the new alliance of nations showed deep down that they respected and supported the interest of slave owners in the North Atlantic colonial empires. This was very contrary to what Bolívar had promised in order to receive the help from Alexandre Pétion in 1815.

The blockade adopted by these empires and its tacit support by the Bolivarian nations lasted until 1865, leaving in complete ruin the vision of a black empire for the diasporas emancipated from slavery in the Caribbean. With the "*fear of another Haiti*", it became clear that the ideas of the revolutions of that time embodied a transition of dominance, from the North Atlantic European monarchies to their liberal, mercantile and bourgeois military elites. The military domination that the creole generals just imposed on their new nations went through maintaining the racist hegemony. By perpetuating its white hegemony in the nascent republics, they paved the way for a consensus of practices of economic and social exploitation on both sides of the Atlantic. This consensus at the end turned into the solid ground to raise the institutions of modern capitalism. The dominance of Imperialism was contested, but the hegemony of racism become secured, in the liberal regimes that presumably were free from brutal slavery.

The fear of "another Venezuela", in 1903

Let's move 100 years ahead. The US imperial state was already beginning to outline its continental hegemony during the Panama conference of 1826. Then the nascent slave empire was still at a disadvantage with Greater Colombia. But while Bolivar proposed a great Colombia from the Andean depths to all corners of the Caribbean, the American Union proposed to expand to the south, even at the risk to confront the Bolivarian forces. Internal struggles between Creole generals plagued the continent with small republics separated from each other. By the end of the century the USA consolidates an expansive policy based on its own image of being a 'superior nation'. With this predicament the US emerged victorious from the wars in the Pacific and the Wars in the Caribbean to annex half of Mexico, defeating the last strongholds of Spain securing control over Cuba, Haiti, Puerto Rico and Panama (Gramsci, 1981a, p. 2017, 220).

After consolidating continental military domination, corporate power emerged as the main force behind the US empire, in building the commercial domination of the production of raw materials from the Iberoamerican republics. Within the countries of the region, Venezuela began to get the attention of corporate colonialism in helping the urban transition in the US through asphalt supply. Benjamin Coates (2015) writes that initially this material began to be exported thanks to a concession granted to the Irish-American Horatio R. Hamilton. A baked-goods purveyor, Hamilton arrived in Caracas as a representative of Vanderveer & Holmes. In 1883 Hamilton received a 25-year concession to exploit forest resources and asphalt throughout the state of Bermúdez. The main attraction of the concession was the Guanoco Lake, some thousand acres in size and eight to ten feet deep full of Bitumen. Since the 1870s engineers had begun mixing bitumen with sand and clay to build asphalt as a durable paving surface (Coates, 2015, p. 385). By 1880 asphalt had become the new chick material to leave behind the dusty past of pre-industrialization and pre-urbanization. The vast open-pit bitumen lakes were bewildering ease to extract and export the new material. Hence, in just a few years, Venezuela became the most coveted country on the continent. By 1885 Hamilton sold his concession to the New York & Brothers Company, which soon established a monopoly of 90 percent of the asphalt supply in the US. Its dominant position generated shortages induced to increase the price of asphalt. The situation generated the appearance of new players in the market, such as the Warner-Quinlan Company, which in 1900 began to demand a part of the Guanoco Lake that was not included in the NY&B concession (Coates, 2015, p. 386).

According to Coates (2015, p. 386), this controversy was induced by the Venezuelan president Cipriano Castro, with the aim of obtaining the assets of

NY&B at a low price. With the certainty of Castro's maneuvers, the NY&B sought support in Washington for a military intervention. But then Roosevelt was not interested in a conflict with Venezuela, although he sent three ships to Venezuela to calm things down. Given this position, NY&B decided to over-throw the Venezuelan president, taking advantage of the conflict that Cipriano Castro had created with Manuel Antonio Matos, then the richest man in Venezuela. Matos had refused to lend money to the Venezuelan state, for which Castro had put him in jail. Instead Matos used his money to buy an artillery ship, El Libertador, which with the support of NY&B was used to start an insurrec-tion against Castro. The NY&B provided ammunition, food and shelter to the soldiers of Matos. As a result of this war more than 15,000 persons died. Cipriano Castro succeeded in defeating Matos and the NY&B army (Coates, 2015, p. 387). But the destructions generated by such corporate colonialism put Venezuela in debt with several other European companies.

As a result of this war Venezuela owed about sixty-two million Bolivares to the German bank *Diskonto Gesellschaft* of Berlin who financed the Great Venezuelan Rail Way Company since 1883 (Maass, 2009, p. 386). This bank headed an impatient consortium pushing the governments of Great Britain and Germany into action. Since Venezuela had no money to pay Germany announced to bomb the country. The conclusion drawn by Roosevelt then was that Venezuela could offer nothing to Germany but territory as indemnity (Morris, 2002, p. 75). Germany has a shortage of living space by side of having the most formidable army in the world by 1902. Morris stresses that as Venezuela turned then into the perfect colony Roosevelt announced "Naval maneuvers on a large scale" to the same theater as the Anglo-German Blockade announced to be held, in November 1902. Being the main source of asphalt to the US as referred by Coates (2015), the situation in Venezuela motivated the US congress to increase the tonnage of built ships to the almost highest level in the world after Britain (Morris, 2002, p. 77). Morris sustains further that the new decision provided the opportunity to implement the Monroe Doctrine in its full. The press played a cardinal role for Roosevelt's moves looking for the support of the press and the public opinion. Knowing the historic property of blockades as a pre-move to an invasion Roosevelt send 53 warships to the scene to show the strength toward the 29 Anglo-German vessels displaced to Venezuela. As Morris (2002, p. 80) writes further, in conversation with the German ambassador Roosevelt made clear that the US was ready to cooperate with the Germans, to find a solution that could gain both parts.

Germany and England asked the United States for permission to occupy Venezuela with the sole objective of collecting their debts. Morris sustains fur-ther that Roosevelt was doubtful about this situation, but sent his army to see that the Europeans did not dare to stay. By 15 December 1902 German-British forces quickly seized, disabled or sunk ten Venezuelan warships. This forced Cipriano Castro to arrest British and German nationals. But the US minister for Venezuela, also in charge of German and British interest, secured their release

(Maass, 2009, p. 387). However, despite their domination at sea, the European empires failed to show their domination on land, which forced them to request the participation of the US to mediate in the collection of the debts. For this purpose, the US appointed Herbert Bowen, its minister for Venezuela, as the representative of the Caribbean nation. In these negotiations, Herbert Bowen proposed that 30 percent of the income of custom houses at la Guaira and Puerto Cabello would be set aside exclusively for the payment to the foreign claimants. He suggested that Belgian officials would be put in charge of administering these customs houses until the entire foreign debt had been paid (Maass, 2009, p. 390). In addition to England and Germany, corporates in other countries like The US, Mexico, Italy, Belgium, France, The Netherlands, Norway and Sweden also demanded compensation (Maass, 2009, p. 392)

With this agreement, for the damages caused during the civil war funded by Matos and the NY&B, the foundations of the Roosevelt corollary were laid out of 'the fear of another Venezuela' that would emerge forcing European powers to launch similar military incursions in the Americas. Since then the USA assumed the role of continental police to intervene for European powers in countries that were at risk of abandoning or contesting the contract of oppression enshrined in the trade terms imposed by legislation adopted in the North Atlantic powers. The Roosevelt corollary laid then the groundwork for a colonial consensus protecting corporate interests in between the empires of the North Atlantic to secure both their commercial, military and cultural hegemony in the Americas and the US commitment to decline to intervene in European internal affairs. In keeping this consensus, European powers have been muted when the US has advanced invasions in almost all the countries of the region there commercial domination has been contested by subaltern forces with in the countries.

However, the corporate and military domination imposed with the Roosevelt corollary has been permanently contested with waves of socialist and nationalist insurrections since the 1950s. With the Cuban revolution from 1959, and the Venezuelan Bolivarian revolution from 1998, the North Atlantic hegemony has faced powerful contestation. Both revolutions, blamed as the rise of Castro-Chavismo, have forced the US empire to use all its tools of economic, media and military domination to coin the fear that 'another Cuba' or 'another Venezuela' will infect the rest of the global south, as we will see in the next two sections.

The fear of "another Cuba"

With the blockade on Haiti, enslaved sugar production shifted to Cuba, with which the island replaced Haiti as the epicenter of the slave trade and labor in the Caribbean. Once again, the intensification of slave exploitation laid the foundations for an emancipatory process also in Cuba. After the victory over the Spanish empire in 1898, the US remained stationed in Cuba and turned the island first into a protectorate and later into a neo-colony (Domínguez López and Yaffe, 2017). US corporations quickly assigned the island the role of a casino and the epicenter of sexual exploitation in brothels and cabarets and the consumption of alcohol and cocaine. In line with these roles in the early postwar years, Havana became the Caribbean capital of sin (Gootenber, 2007, p. 150).

"And then Fidel arrived", as it sounds in the emblematic song of Carlos Puebla. The Cuban Revolution completely broke with American cultural hegemony. The processes of expropriations of land, properties and factories initiated by the revolution exacerbated the confrontation with the privileged oligarchies that with the cocaine mafias moved to Miami, New York and Los Angeles. Initially, the US diplomacy wanted to accuse Cuba of being a node of cocaine consumption to demonize the government of Fidel Castro. However, it soon became clear that the one who really had become a bandit state was the US. This bandit diplomacy allowed the privileged Cuban anti-communist diaspora to maintain control of the distribution of cocaine in the main urban centers of the country, and at the same time act as the head of terrorist actions to destroy the remnants of the Cuban tourist industry. The military domination and terrorism of the US increased, while the European powers were busy with their reconstruction and loyal to the agreement in the Roosevelt corollary. In this context, Fidel Castro surprised J.F. Kennedy when Cuba got the support of the Soviet Union to install atomic missiles aimed to repel an eventual US invasion (Brenner, 1990). From then on, the domination of the North Atlantic powers shifted character to the economic, media and technological blockade, or the war-and-repression-schema of domination in Foucault's vocabulary, that has lasted until today.

Double in population size and three times the land area of Haiti, the Cuban revolution since 1959 has continued the anti-racist and anti-capitalist struggle assisting the emancipation of the countries of the global south. The Cuban counter-hegemony have been very little studied by the academic world, to approach a response to the North Atlantic powers domination, I will depart from the work of Richard Twine, who outlined an analysis of the entanglements between different complexes (industrial, agricultural, pharmaceutical, science, entertainment etc) necessary for global capitalism to secure global hegemony (Twine, 2012). Cuba is the only country in the global south that has succeeded to develop its own military, industrial, educational, pharmaceutical and cultural complexes to respond the North Atlantic elites on a domestic and global scale. Although a more profound analysis would demand another book, I will stress some of these Cuban counter-hegemonic complexes even to stress the decolonial character of the Cuban revolution.

In terms of the agricultural and industrial complexes Cuba directed its sugar production entirely towards the Soviet Union, from where it received materials, oil and machinery for its agricultural and industrial development. In recent years Cuba has developed a successful model of urban agriculture, free of pesticides in open contestation to the chemical industry and its transgenic aberrations (Altieri et al., 1999). This model has been advanced recently as example of a sustainable political strategy for a low-carbon, degrowth agenda (Cederlöf, 2016).

To contest the domination of the military complex, upon defeating the dicta-torship of the dictator Batista and after the missile crisis, Cuba dispatched mili-tary missions and guerilla training in many of the anti-imperialist and anti-racist insurrectionary processes of the global south. Its main victory was achieved in southern Africa during the mid 1980s, contributing to the liberation of Angola and Namibia, which were cardinal to end the hegemony of apartheid in South Africa (Kasrils, 2013).

In terms of the educational complex, the Cuban revolution has imposed its own hegemony in public education with literacy campaigns throughout the global south. Also related is the development of the medical pharmaceutical complex, since the 1980s, Cuba laid the foundations for its own biotechnological industry and medical diplomacy. Since the 1960s the Cuban self-management model has established new counter-hegemonies in medicine, achieving extraor-dinary successes against the most devastating tropical diseases in the global south, from malaria to Ebola (Beldarraín Chaple & Mercer, 2017). With the impressive deployment of medical personnel displaced to some forty countries around the world in the fight against Covid-19 (Fitz, 2020), Cuba has shown the resilience of a health model totally opposed to the neoliberal privatisation of the human right to health care.

In terms of the entertainment complex, to raise dollars, Cuba has made pro-gress in consolidating its tourism industry. In popular culture, the Cuban cul-tural revolution has also imposed its counter-hegemony in different musical genres, especially from progressive trova (Moore, 2003) to Afro-Latin dance music. Beyond these music branches there is also a new cinema of social content, and finally Cuba, despite its enormously limited material resources, remains as the principal Olympic power from the global south in the world (Morton, 2002). All in all, the Cuban revolution has managed to develop counter hegemonic models not only to withstand the strict blockade imposed by the North Atlantic colonial powers, but also to project new anti-imperialist and anti-colonialist counter-hegemonies in the Caribbean region and beyond. The fear for another Cuba is a motive of concern, and the worst scenario is its dissemination for the rest of the continent, which, as will be addressed next, the Bolivarian revolution in Venezuela attempts to achieve.

The fear of 'another Venezuela', again

One of the most central processes breaking up with US hegemony and domi-nation in the Americas is the Bolivarian process. I am not going to delve much into the beginnings and developments of the Bolivarian revolution since there is plenty of literature on its progress and achievements. Instead of insisting on its fabulous counter-hegemonic virtues, I would like to focus on the relationship between hegemony and domination from within the revolution to make clear its implications for decolonizing the Bolivarian process.

The Bolivarian revolution started by Hugo Chávez during the 1990s. He won the presidential elections in 1998, promising to overcome the consensus of neo-liberal governance by reforming the constitution to introduce participatory democracy. A year later the new Bolivarian constitution was adopted aiming to implement the active and leading participation of the people in the management of society. Around 2002 the new political constitution served to unleash a great wave of black, mestizo and indigenous empowerment around the construction of communal councils and the so-called poder popular, citizen power.

The exchange of technical and medical and other personnel from Cuba for Venezuelan oil supplies laid the foundations for a productive counter-hegemonic alliance with the creation of ALBA. The alliance has kept both countries afloat under a model of complementarity that for ten years replaced imperial dominance in the supply of fossil fuels in the small nations of the Caribbean region through Petrocaribe. This collaboration continued towards the creation of an anti-imperialist news network (TeleSur) (Burch, 2007), the creation of a South American body for military and security issues (UNASUR) and another continental body for anticolonial cooperation (CELAC) (Quiliconi and Salgado Espinosa 2017). The late looking for an strategic alliance with China and other powers beyond the North Atlantic powers dominance (Bonilla Soria and Herrera-Vinelli 2020).

When Hugo Chavez died in cancer in 2013 all this work started to falter. In 2015, US President Barack Obama announced an emergency situation to fight Venezuela as an 'unusual' threat to US national security (Koerner, 2015). To implement 'The fear for another Venezuela' Obama introduced sanctions that gave oppositional entrepreneurs, organized in fedecamaras, the courage to hide goods in order to drive up inflation and undermine the government's economic policy. This tactic gave political results, when the opposition won a two-third-part majority in the 2015 parliamentary elections. This victory provided the opposition the opportunity to dismiss ministers and call a constituent assembly to rewrite the constitution. But three members elected for Amazonas state were indicted for buying votes. They acknowledged the crime and gave up their seats in Parliament. New elections were to be held in their district, but the opposition refused to obey both the electoral authority and the Supreme Court, and later declared itself in disobedience to the other state powers. Under these conditions, the Supreme Court ruled that all decisions of Parliament were invalid as long as the Parliament was in disobedience.

The opposition then turned to the streets to overthrow the government with violent riots. To respond to the wave of violence and the opposition's refusal to dialogue, on May day, president Maduro announced the election of a constituent assembly by July 30, 2017. The idea was that that assembly would find a structural solution to the country's problems and deepen the Bolivarian revolution. The opposition countered threats to stop the election by force and demanded Maduro's departure. The elections were held and the violent riots ceased almost immediately after that. The new Constituent Assembly called for the election of governors, passed a new law against hate crimes, and preceded presidential elections under May 2018. However, the US government, now headed by Donald Trump, neglected to recognize all these elections and imposed new tougher financial sanctions that crashed the oil production, access to credit markets, stopped Venezuela to restructure debt, import food and medicines. Altogether resulted in higher mortality rates (Pantoulas and McCoy, 2019, p. 403).

Before the installation of Maduro as president on January 5, 2019, the Parliament elected Juan Guaidó as president to head its sessions during 2019. The conflict with the Parliament in disobedience forced judges in the Supreme Court to step in to swear President Maduro into office. On January 23, in front of a public gathering in Chacao, the country's most gentrified municipality, Juan Guaidó, as head of the Parliament, declared President Maduro insane and self-appointing himself as interim president. Guaidó was immediately backed by the US leadership through a cascade of Twitter posts, which suddenly replaced the official constitutional and diplomatic channels (Vaz, 2019). Since then the US's hegemony in social media corporations has been used to exercise further dominance over Venezuelan politics. About 60 countries, mostly from the European Union and the Americas supported the move adopted by Washington, demanding Russia and China to back off Venezuela. As with Haiti, Venezuela in 1903, and Cuba since 1959 European powers were on the same page as the U.S. This time rejecting universal elections, the UN Charter and international law to reinstate coup d'états.

While this struggle for the domination of Venezuelan resources went on the level of the world system, the ideological struggle for revitalising the hegemony of the bourgeoisie as a historic subject went at the everyday level towards grassroots movements. The National Constituent Assembly that was elected to deepen the Bolivarian revolution, was unable to formulate the new chart and its president, Diosdado Cabello announced its dissolution by 31 december 2020 (Dobson, 2020). By doing so it confirmed to be a cloud of dispersed leaders manipulated by the government's political manoeuvres.

By 2017 Maduro had promised to deepen the alliance with grassroots movements to deepen the hegemony of the proletariat as the subject of the revolution. But ignoring the warnings of these movements Maduro focused on giving right-wing bourgeois constellations a stronger position to participate in the 2020 congressional election. Since 2018, his Minister of Agriculture, Wilmar Castro Soteldo, part of the old nomenclature chosen by Chavez together with Maduro to head the Bolivarian revolution, openly advocated the need to create a 'revolutionary bourgeoisie'. This is a stratum of "new rich" constituted by business-people and civic-military bureaucrats of whom some also become entrepreneurs and landowners (Perez, 2020). According to Castro Soteldo, this revolutionary bourgeoisie is the only one that in practice could success on the task of getting the country's economy back on its feet.

In response to the 'revolutionary bourgeoisie's race for dominance in the government, in August 2020, the Alternativa Popular Revolucionaria, APR, was formed conducted by the Communist party to represent grassroots and small left-wing parties to the election of the National Assembly in December (Pascual Marquina, 2020). Maduro and the PSUV selected their candidates from the same networks that lost the election in 2015, and who built up their dominance over institutions through corruption, nepotism and threats. In the APR appeal, many saw a brightening to elect a more grassroots parliament to make demands on Maduro's government, like the right to a salary to keep basic life standards, as stated by the Bolivarian constitution, while safeguarding the achievements of the revolution and keeping an anti-Imperialist position. Maduro's and the PSUV's response was to stigmatize and diminish these parties and movements and to legitimize the imprisonment and deprivation of their representatives' democratic rights by the Supreme Court. To make this regime shift even more evident, Maduro on the one hand cleared left-wing alternatives to the PSUV in early September 2020, and on the other hand issued a decree pardoning 110 of the coup plotters who committed countless terrorist crimes (Solé, 2020). To secure a stronger opposition to face the imperial blockade imposed by the North Atlantic empires Maduro made big efforts to reach the participation of a wide spectrum of bourgeoise parties to restablish the legitimacy of the Venezuelan parliament, Asamblea Nacional, and so the legitimacy of the bolivarian state in the international community.

However, learning from Gramsci's (1981b, p. 172) analysis, although the communal councils achieved intellectual political hegemony in civil society during the time of the Chávez government, they have experienced fierce opposition to secure such dominance in the political society. As explained, during the Maduro government this dominance, in the hands of the United Socialist Party, PSUV, empowered the 'revolutionary bourgoisie', as an instrument of coercive and instrumental domination within the state. With this move, presumably adopted as an anti-imperial maneuver, the government of Maduro aimed to get the support of the international community of liberal states.

Like in the case with Haiti, once again, the anti-imperialist interests were dominating, while the interest of the subaltern black and brown proletariat were subordinated to the hegemony of the local and global bourgeoisie.

Concluding remarks

As has been explained, in the period previous to the formulation of the Monroe doctrine a black hegemony was instituted to abolish slavery with the Haitian revolution. But this hegemony was countered both by the colonial and the anti-imperialist elites that coined the *fear for another Haiti* to secure white dominance even among the post-imperial creole states. With the consolidation of the Monroe doctrine, this fear for another anti-imperialist moment was functional to frame the *fear for another Venezuela* in 1903. Then the North Atlantic empires agreed the Roosevelt corollary in the Monroe Doctrine to secure domination over the Americas in the years to come. In 1959 the Cuban revolution began to establish a socialist hegemony that flooded beyond the Caribbean over to the global south. Its strategic alliance with Venezuela has been very cardinal to navigate further to contest the Monroe doctrine. In Venezuela a proletarian hegemony could be seen securing the dominance of grassroots democracy from 1998 to 2013. But since then *"The fear for another Venezuela"* seems to succeed with a regime shift, this time headed by those who surfed on Chavez's legacy, and that headed by the 'revolutionary bourgeoisie' excluded and banned the proletariat class, their representatives and organisations, from heading the state affairs.

The Venezuelan case shows that in order to secure the hegemony over the black and colored proletariat the domestic bourgeoisie agreed an historical consensus to transit from subjugation to the Monroe doctrine of the North Atlantic empires towards a new era of alliances with the 'revolutionary bourgeoisies' from China, Russia, Turkey and Iran. It is hard to believe that this would mean the end of the Monroe doctrine and the domination the North Atlantic bourgeoise empires. Instead, this can frame the resurgence of the hegemony of a much more diverse global bourgeoisie in a time when its racism and environmental practices continues to head planetary destruction.

References

Altieri, M., Companioni, N., Cañizares, K., Murphy, C., Rosset, P., Bourque, M. and Nicholls, C. (1999). The greening of the "barrios": Urban agriculture for food security in Cuba. *Agriculture and Human Values, 16*(2), pp. 132–140.

Beauvois, F. (2010). Monnayer l'incalculable? L'indemnité de Saint-domingue, entre aproximations et bricolage. *Reveu Historique, CCCXII*(3), pp. 609–635.

Beldarraín Chaple, E. and Mercer, M. A. (2017). The Cuban response to the ebola epidemic in West Africa: Lessons in solidarity. *International Journal of Health Services, 47*(1), pp. 134–149.

Bonilla Soria, A. and Herrera-Vinelli, L. (2020). CELAC como vehículo estratégico de relacionamiento de China hacia américa Latina (2011-2018). *Revista CIDOB d'Afers Internacionals*(Nr. 124), pp. 173–198.

Brenner, P. (1990). Cuba and the missile crisis. *Journal of Latin American Studies, 22*(1–2), pp. 115–142.

Burch, S. (2007). Telesur and the new agenda for latin American integration. *Global Media and Communication, 3*(2), pp. 227–232.

Cederlöf, G. (2016). Low-carbon food supply: The ecological geography of Cuban urban agriculture and agroecological Theory. *Agric Hum Values, 33*(4), pp. 771–784.

Chambers, S. (2006). Masculine virtues and feminine passions: Gender and Race in the republicanism of simon Bolivar. *Hispanic Research Journal, 7*(1), pp. 21–40.

Coates, B. (2015). Securing hegemony through law: Venezuela, the U.S. Asphalt trust, and the uses of International law, 1904-1909. *The Journal of American History, 102*(2), pp. 380–405.

de la Reza, G. (2015, Septiembre). El intento de integración de Santo Domingo a la Gran Colombia (1821-1822). *Secuencia*(93), pp. 65–82.

de Melo Rosa, R. and Pongnon, V. N. (2013). A república do Haiti e o processo de construção do estado-nação. *Revista Brasileira do Caribe, XIII*(26), pp. 461–494.

Dobson, P. (2020, September 7). *Venezuela's Constituent Assembly: No New Constitution on the Way*. Retrieved September, 27 2020, from Venezuelanalysis: https://venezuelanalysis.com/YwwH

Domínguez López, E. and Yaffe, H. (2017). The deep, historical roots of Cuban anti-imperialism. *Third World Quatterly, 38*(11), pp. 2517–2535.

Fanon, F. (1967). *The Wretched of the Earth*. London: Penguin books.

Femia, J. (1975). Hegemony and consciousness in the thought of Antonio gramsci. *Political Studies, 23*(1), pp. 29–48.

Ferrer, A. (2012). Haiti, free soil, and antislavery in the revolutionary Atlantic. *American Historical Review, 117*(1), pp. 40–66.

Fitz, D. (2020). How che guevara taught Cuba to confront COVID-19. *Monthly Review, 72*(2), pp. 21–30.

Foucault, M. (2003). *Society must be defended: Lectures at the college de France, 1975-1976*. New York: Picador.

Gootenber, P. (2007). The "Pre-Colombian" era of drug trafficking in the Americas: Cocaine, 1945–1965. *The Americas, 64*(2), pp. 133–176.

Gorender, J. (2004). O épico e o trágico na história do Haiti. *Estudos Avancados*, 18(*50*), pp. 295–302.

Gramsci, A. (1981a). Cuaderno 1 (XVI) 1929-1930. In V. Gerratana (Ed.), *Cuandernos de la Carcel – Tomo I*(A. M. Palos, Trans. pp. 73–196). Mexico DF: Ediciones Era.

Gramsci, A. (1981b). Cuaderno 4 (XIII) 1930–1932. In V. Gerratana (Ed.), *Cuadernos De La Carcel - Tomo II* (A. M. Palos, Trans., pp. 129–244). Mexico DF: Ediciones Era.

Kasrils, R. (2013). Cuito Cuanavale, Angola: 25th anniversary of a historic African Battle. *Monthly Review, 64*(11), pp. 44–51.

Koerner, L. (2015, March 9). *U.S. President Barack Obama Brands Venezuela a "Security Threat," Implements New Sanctions*. Retrieved September, 27 2020, from Venezuelanalysis: https://venezuelanalysis.com/NoZL

Maass, M. (2009). Catalyst for the roosevelt corolary: Arbitring the 1920-1903 Venezuelan crisis and its impact on the development of the roosevelt corollary to the Monroe doctrine. *Diplomacy & Statecraft, 20*(3), pp. 383–402.

Monguey, V. (2012). A tale of Two brothers: Haiti's other revolutions. *The Americas, 69*(1), pp. 37–60.

Moore, R. (2003). Transformations in Cuban Nueva trova, 1965–95. *Ethnomusicology, 47*(1), pp. 1–41.

Morris, E. (2002). A matter of extrem urgency": Theodore Roosvelt, Wilhem II, and the Venezuelan crisis of 1902. *Naval War College Review, 55*(2, Article 6), pp. 1–13.

Morton, H. (2002). Who won the Sydney 2000 olympics?: An allometric approach. *Journal of the Royal Statistical Society. Series D (The Statistician), 51*(2), pp. 147–155.

Pantoulas, D. and McCoy, J. (2019). Venezuela: An unstable equilibrium. *Revista de Ciencia Politica, 39*(2), pp. 391–408.

Pascual Marquina, C. (2020, September 11). *Pursuing National Liberation and Socialism: A Conversation with Oscar Figuera*. Retrieved September 27 2020, from Venezuelanalysis: https://venezuelanalysis.com/YwwF

Perez, L. (2020, 6 24). *The Venezuelan New Rich's Taste for Artwork*. Retrieved September 27 2020, from Venezuelanalysis: https://venezuelanalysis.com/Yw3p

Quiliconi, C. and Salgado Espinosa, R. (2017). Latin American integration: Regionalism à la carte in a multipolar world? *Revista Colombia Internacional*(Nr.92), pp. 15–41.

Romero Jaramillo, D. (2003). El fantasma de la revolución haitiana esclavitud y libertad en Cartagena de indias 1812–1815. *Historia Caribe, III*(8), pp. 19–33.

Shah, K. (2009). The failure of State building and the promise of State failure: Reinterpreting the security- development nexus in Haiti. *Third World Quarterly, 30*(1), pp. 17–34.

Solé, M. (2020, September 2). *Venezuela's Maduro Grants Pardon to 110 Opposition Figures.* Retrieved September 27, 2020, from Venezuelanalysis: https://venezuelanalysis.com/YwiQ

Twine, R. (2012). Revealing the 'animal-industrial Complex' – A concept & method for critical animal studies? *Journal of Critical Animal Studies, 10*(1), pp. 12–39.

Vaz, R. (2019, January 23). *Venezuelan Opposition Leader Guaido Declares Himself President, Recognized by US and Allies.* Retrieved September 27 2020, from Venezuelanalysis: https://venezuelanalysis.com/N4H5

13

INTERNATIONAL DEVELOPMENT FINANCING IN A POST-BRETTON WOODS WORLD

Syed Sajjadur Rahman

Introduction

Whereas the 1980s and 1990s were marked by an accelerated growth of globalization, the current epoch is noticeable for the chaotic retreat of the North[1] and the disintegration of neoliberal alliance formed immediately after the Second World War (WWII).

The "neo-liberal" term has been widely interpreted (Ritzer 2010, Ch. 5). In its widest form, it is interpreted as "free markets" and "small governments". The victorious Northern allies had a more material interpretation – to countervail the Soviet dominated spread of communism and to prolong the dominance of colonial states on their ex-colonies or soon to be ex-colonies. All this was to take place under the strict supervision of the North through their control of the decision-making in the international institutions set up at the 1944 Bretton Woods Conference, stage managed by the US and its principal ally, the United Kingdom (UK).

The Bretton Woods Institutions (BWIs) - the International Monetary Fund (IMF), the International Bank for Reconstruction and Development (later the World Bank), the still-born International Trade Organization (later reformulated as General Agreement Tariffs and Trade (GATT) and subsequently World Trade Organization (WTO) - were set up to implement the intents of their founding nations. Here the term BWIs is used in a broader sense and includes the regional development banks (such as the African, Asian and Inter-American Development Banks) as well as northern bilateral development assistance agencies, because they were all part of the same hegemonic construct serving Northern interests and objectives.

The provision of international development financing was dominated by the BWIs from 1945 to the end of the Twentieth Century. This domination is now in rapid decline. Alternative financial sources have emerged. The South[2] is increasingly able to finance its own development and the North is in an isolationist retrenchment mode. It seems that the Covid-19 pandemic will further reduce the influence of the North.

Section II briefly traces the evolution and performance of the BWIs. Section III chronicles the rise of the South and Section IV analyses the impact of this rise in terms of international development finance. Section V concludes the Chapter with a brief discussion on the future of international development financing and the possible impact of the Covid-19 pandemic.

Bretton Woods Institutions – origins and performance

The origins of the Bretton Woods Institutions (BWI)

The Bretton Woods Conference (formally, the "The United Nations Monetary and Financial Conference") was held in July 1944. Its purpose was to "plan a new economic order to the post-war world which would avoid a repeat of the disastrous policy mistakes of the 1920s and 1930s" (Pickford 2019). There were 44 participants - there was even one participating entity called the British Raj (India).

Three factors influenced the deliberations and decisions at this Conference.

- Avoiding economic conflicts between nations by promoting international norms and establishing institutions to help monitor and adjudicate their implementation: The participants adopted a neoliberal philosophy. The IMF was to monitor and regulate fiscal and monetary affairs; the ITO/GATT[3], the trade matters and the IBRD (World Bank) was to help rebuild countries devastated by World War II, but over time to "foster" international development.
- Creating a sphere of influence to combat the growing "spectre" of communism: The BWIs were to provide the central fulcrum of this sphere providing the normative and the practical expression of benefits of belonging to the "free world". The Soviet Union and other communist countries eventually became members but, the institutions were designed to remain in firm control of the North.
- Preserving the North's colonial hegemony over the South: The European colonial powers were considerably diminished in their ability to maintain control over their colonial empires in the South post WWII. The strategic question facing them was how they could retain control of the colonies without incurring large military expenditures. The BWIs offered ideal venues for achieving these objectives. They exerted Northern dominance based on a notion of "superiority" in managing a world order based on a Northern value-based "rules" oriented international economic order enforced by the overwhelming military and financial power of the US. The benefits offered to ex-colonies were access to northern markets, foreign direct investment, becoming part of the global supply chains, and aid.

The IMF and the IBRD (World Bank) were formally established in March 1946. A watered-down version of the ITO called the GATT was ratified and established on 1948. The regional development banks (RDBs) were established much later and were modeled closely after the World Bank – the Inter-American Development bank (IADB) was set up in 1959, the African Development Bank (AfDB) in 1964, and the Asian Development Bank (ADB) in 1966. Most Northern bilateral aid agencies were set up in the 1960s and 70s.

Performance

Trade Liberalization

Multilateral and bilateral/regional agreements were the main instruments for liberalizing trade.

Multilateral Agreements: There have so far been nine multilateral trade rounds resulting in significant liberalization of trade (Table 13.1), the first eight under the auspices of GATT. The ninth (Doha) Round launched in 2001 with an explicit focus on addressing the needs of the developing countries has not yet concluded, with major disagreements between the North and the South on freer trade in industrial goods and services, continuation of agricultural subsidies, and treatment of intellectual property rights.

Bilateral or Regional trading Arrangements: These agreements widened markets, deepened integration and promoted economies of scale for the participants. However, they were inherently discriminatory as they diverted trade and deprived excluded countries (most often the weak states from the South) of income.

Overall, the multilateral trade liberalizations and the resultant globalized production processes have economically benefitted the developing regions (Table 13.1). But, the distribution of the dividends of globalization has been biased towards the North.

Poverty reduction

In September 2019, the World Poverty Clock (World Data lab 2019) estimated that 8 percent of the world's population was living on less than $1.90 a day (*extreme poverty*), down from nearly 36 percent in 1990 (World Bank, 2018). Extreme poverty is now becoming concentrated in Sub-Saharan African (World Bank, 2018).

Poverty reduction impacts of the BWIs are unclear and difficult to disentangle because aid accounts for a relatively small share of financial flows to the South. Aid seems to have a positive impact on growth in countries with good fiscal, monetary, and trade policies but has little effect in the presence of poor policies (Burnside and Dollar, 2000). Where political institutions are weak, aid transfers are vulnerable to expropriation and may lead to capital flight, undermine growth and strengthen the hand of the political elite. Nevertheless, Easterly (2003) concluded that, "The goal of having the high-income people make some kind of transfer to very poor people remains worthy one despite disappointments of the past". But the goal needs to be modest.

TABLE 13.1 GATT and the WTO Trade Rounds

Name	Start	Duration	Countries	Issues	Results
Geneva	1946	7 months	23	Tariffs	Signing of GATT, 45,000 tariff concessions affecting $10 billion in trade
Annecy	1949	5 months	13	Tariffs	Countries exchanged 5000 tariff concessions
Torquay	1950	8 months	38	Tariffs	Countries exchanged 8700 tariff concessions, cutting 1948 tariff levels by 25%
Geneva II	1956	5 months	26	Tariffs, admission of Japan	$2.5 billion in tariff reductions
Dillon	1960	11 months	26	Tariffs	$4.9 billion in tariff concessions
Kennedy	1964	37 months	62	Tariffs, anti-dumping	$40 billion in tariff concessions
Tokyo	1973	74 months	102	Tariffs, non-tariff measures, framework agreements	$300 billion in tariff reductions
Uruguay	1986	87 months	123	Tariff, non-tariff measures, rules, services, intellectual property, dispute settlement, textiles, agriculture, creation of WTO	Creation of WTO, about 40% reductions in tariffs, major reductions in agricultural subsidies, agreement to allow full access for textiles and clothing from developing countries, extension of intellectual property rights
Doha (WTO)	2001		141	Tariffs, non-tariff measures, agriculture, labour standards, environment, competition, investment, patents, transparency	The Round is not yet concluded.

Source: Various GATT and WTO documents

Perpetuation of Colonial Influence

Alesina and Dollar (2000) found that for bilateral donors, the pattern of aid giving is dictated by political and strategic considerations – "an inefficient economically closed, mismanaged non-democratic former colony politically friendly to its former colonizer, receives more foreign aid than another country with similar level of poverty, a superior policy stance but without a past as a colony"[4]. Aid has also "historically been seen as an instrument of United States and other Western countries' political and economic power" (The Bretton Woods Project, 2019).

The governance structures of the BWIs perpetuated the domination by the North. At the IMF and the World Bank, decisions are in principle taken based on a voting system where the voting rights are based on the member countries share of the world's GDP. However, actual practice indicates that while the OECD countries voting share at the IMF in 2016 was 63.09 percent, its share of the world economy was 45.60 percent (Weisbrot and Johnston, 2016). The two most obviously underrepresented countries were China (voting share 6.16%, share of world economy 18.59%) and India (2.67, 7.09) (Ibid: 2016). Table 13.2 provides the shareholder statuses at the IBRD and the International Development Association – the grant giving arm of the World Bank. Once again, the Northern countries enjoyed the dominant voting power. This unequal balance of power is reinforced by a "gentleman's agreement" between the US and the European Countries that the head of the World Bank come from the US and that of the IMF from Europe.

This type of shareholder prerogative could be appropriate for a private entity. But, to the extent the BWIs are also "development" institutions, the other stakeholders in the development process principally the Southern countries were only marginally represented in decisions that mainly affected them. This system perpetuated the donor-recipient, metropolis-satellite (Frank, 1966), colonizer-colonized, and lender-creditor relationships where the first party in the relationships was the dominant entity.

The impact of this decision-making process was evident in the policies and actions of these institutions. The policies were dominated by neo-liberal

TABLE 13.2 IBRD and IDA principal shareholders 2009

Country	IBRD Share (%)	IDA Share (%)
US	16.38	12.15
Japan	7.86	9.64
Germany	4.49	6.38
France	4.30	5.03
UK	4.30	5.03
China	2.78	1.91
India	2.78	3.17

Source: Global Policy forum (2009)

principles embodied later in the Washington Consensus including the much-discredited structural adjustment policies favoring draconian reductions in public services and "monetary discipline" that adversely affected the poor. The BWIs were also marked by biased and inconsistent decision making when Northern interests were at stake (Bretton Woods project, 2019). Their actions often led to violations of human rights and to environmental damages caused by projects that promoted economic growth at the expense of sustainability (Ibid: 2019). And all of this took place in an international context where despite an US Supreme Court decision of culpability, the BWIs operated with impunity and without any fear about the consequences of their actions (Ibid: 2019).

The rise of the South: Evidence and perceptions

Radical changes are now taking place in the global hegemonic construct. There has been an attenuation of the North's hold on global economic and strategic circumstances that has diluted the ability of the BWIs to influence the course of international development efforts.

The rise of the South

Ironically, this attenuation owes its origin and progress to the very mechanisms that were meant to preserve the Northern hegemony like liberalization of trade and FDI, the encouragement of pluralistic governance systems and aid. While economic globalization in the form of global supply chains allowed multinational corporations (MNCs) to take advantage of country comparative advantages, it also provided the South a foothold in the global markets. Many countries implemented export-oriented industrialization processes to profit from increased access to Northern markets.

These mechanisms sparked the economic rise of the South and the emergence of regional powers like China, India and Brazil (Table 13.3). South-east Asian

TABLE 13.3 GDP Growth (annual %) in Emerging Economies

Region/Country	1980–89	1990–99	2000–2009	2010–18	Classification by Income
Africa					
South Africa	2.24	1.39	3.66	1.85	High MIC
Americas					
Brazil	2.99	1.69	3.3	1.42	High MIC
Colombia	3.41	2.86	4.13	3.76	High MIC
Asia					
China	10.1	9.64	10.23	7.79	High MIC
India	5.69	5.79	6.9	7.02	Low MIC
Indonesia	6.39	4.83	5.1	5.46	Low MIC
Malaysia	5.89	7.23	4.79	5.43	High MIC

Source: World Bank, https://data.worldbank.org/indicator/NY.GDP.MKTP.KD.ZG?locations=TH

TABLE 13.4 HDIs and Inequality in Emerging Economies

Country	HDI Rank 2018	1980	2000	2010	2018	Inequality adjusted HDI 2018	Gini Coefficient 2010–2017
Africa							
South Africa	113	.569	.628	.638	.705	.463	63.0
Americas							
Brazil	79	.545	.682	.739	.761	.574	53.3
Colombia	79	.557	.655	.706	.761	.585	49.7
Asia							
China	85	.423	.591	.681	.758	.636	38.6
India	135	.369	.483	.570	.647	.477	35.7
Indonesia	111	.470	.609	.671	.707	.584	38.1
Malaysia	61	.577	.717	.716	.804	–	41.0

Source: UNDP, Various Human Development Reports

countries reaped the highest benefits of the trade liberalization. Overall, the developing world is now more "middle-income" (111 countries in 2019) than "low-income" (34)[5]. This is not to say that income is the only measure of development. Rather, income is an indicator of economic ability. The higher the income level, the more a country is able to finance its own development efforts.

Many of the Southern countries have now become high human development countries with human development indexes over 0.7 (Table 13.4). However, growth dividends can be unevenly distributed leading to unequal development. Table 13.4 provides data on inequality adjusted HDI as well as the Gini-coefficient. On average, the HDI diminishes by about a quarter of its original value once inequality is factored in. The Gini coefficients indicate that inequality remains a pervasive problem especially in Africa and the Americas.

Nevertheless, the result is an increasingly assertive South whose development aspirations are more nationally driven than externally dictated. For example, the MDGs were often used as development goals in national development plans. In contrast, the SDGs are now used more as indicators rather than as primary goals in many planning frameworks (Zaman and Rahman, 2020). There has also been open expressions of distrust and disavowal of Northern policy directions including disagreements on how to deal with the financial crisis of the late 1990s and the global recession in 2008 precipitated by the collapse of the US housing and financing bubbles.

The rise of the South has been marked by the rise of potential regional hegemons. The most obvious example is China which has emerged as not only the preeminent power in Southeast Asia but as the second largest global economy poised to take over the first position relatively soon. China is now considered the manufacturing hub of the world, a dependency highlighted during the recent Covid-19 pandemic. Others include Brazil and India.

The attenuation of Northern influence and the Northern reaction: The anti-globalization movements and its political expressions

The US and many other Northern countries are retreating from their global hegemonic roles into ultra-nationalist and isolationist shells. As Lind and Press (2020) put it, "In the United States and among several of its core allies, large parts of the public have lost confidence in the liberal project that long animated Western foreign policy. The disillusion is in part a reaction to the twin forces of economic globalization and automation which have decimated employment in manufacturing in the developed world. It is also reflected in the growing opposition to immigration, which contributed to the United Kingdom's vote to leave the EU, the rise of chauvinist parties across Europe, and the election of Donald Trump in the United States". These anti-globalization movements in the North have been sparked by calculation (differential perceived costs and benefits of integration – the rust-belt syndrome), community fears (the "losers" subscribing to a more exclusive national identity and anxiety over the perceived negative effects of immigration), and cues (images of opinion leaders with partisan attachments and ideological predispositions) (Hooghe and Marks, 2005).

For the US, the dominant force in the Northern neoliberal coalition, one reaction has been a move away from multilateralism. It has withdrawn from the Trans-Pacific Partnership and is planning to withdraw from the World Health Organization. It has also engaged in predatory tariff wars with China and others. One could interpret these actions as hostile reactions to the loss of its unilateral global leadership status – "US leaders felt free to reimagine reality largely unconstrained by the objections of those who opposed the global liberal project" (Lind and Press, 2020). But the reality is, "In 1970, the West produced 56 percent of the world economic output and Asia (including) Japan produced only 19 percent. Today, only three generations later, those proportions have shifted to 37 percent and 43 percent – thanks in part to the staggering economic growth of countries such as China and India" (Milanovic, 2019).

From a civil society perspective, the emphasis on development cooperation in the North has all but disappeared. There has been a "financialization" of aid, partly in response to the overwhelmingly middle-income nature of the developing regions. Aid is being turned into a de-risking and leveraging instrument. The impact has been a diminution of the roles of Northern NGOs and CSOs.

Implications for International development financing

The lackluster performance of the BWIs in advancing international development as well as the rise of the South are starting to transform how international development is financed.

The major portion of development funding generally comes from the countries themselves. Between 2000 - 2015, developing countries have increased domestic revenues by an annual average of 14% - the domestic revenues of developing economies amounted to US$7.7 trillion in 2012, US$6 trillion more than in 2000 (UNDP, 2015). Important challenges remain. In the least developed countries, tax revenues amount to just 13% of GDP on average - about half the level in many other developing countries (UNDP, 2015). Nevertheless, as more countries become MICs, their ability to generate domestic revenues continue to increase diminishing the need for external funds.

The external financing sources include international trade, foreign direct investment and other private flows including remittances, aid and external debt.

International trade

International trade increased by more than 50% percent between 2005 - 2015, with about 60% of the increase tied to rising exports from developing countries (UNCTAD, 2016). South-South trade has grown even faster over the same period - more than tripling to reach 57% of all developing country exports (US$9.3 trillion) in 2014 (UNCTAD, 2016). The rapid growth in South–South trade reflects a lessening of dependency on Northern markets. The labor-intensive production stages of many industries have relocated from countries like China and India to lower-wage economies (Subramanian and Kessler, 2013). A host of new institutions such as China's Belt and Road Initiative (BRI) and the Asian Infrastructure Investment Bank (AIIB) are supporting South-South trade and investments. Overall, many countries have gone from being aid dependent to trade reliant over the last three decades.

FDI and remittances

In 2016, more than 40 percent of the nearly $1.75 trillion of global FDI flows was directed to developing countries (World Bank, 2018a). FDI by Southern firms accounted for nearly one-fifth of global FDI flows in 2015, up from just 4 percent in 1995. The BRICS (Brazil, Russia, India, China, South Africa) investors are the key drivers, accounting for 62 percent of total developing country FDI stock in 2015 – with China alone accounting for 36 percent (ibid (2018a). In 2015-16, the ten leading foreign investors in Africa, by number of new projects, included China, India, Kenya and South Africa (The Economist, 2018). While both Northern and Southern investors respond to the same type of locational determinants (market size, income level, distance, common language, colonial links), Southern investors are more willing to target smaller and closer economies (World bank, 2018a). In addition, managers of developing country MNCs may be more accustomed to uncertainty and more adept in dealing with unpredictable regulatory practices and less transparent administrative procedures (World Bank, 2018a)

Remittances: Remittance inflows reached US$529 billion in 2018. India ($79 bn, 2.9% of GDP) and China ($67 bn, 0.4%) were the top recipients followed by Mexico ($36 bn, 3%), The Philippines ($34 bn, 10.2%), and Egypt ($29 bn, 11.6%). The official remittance channels remain expensive (global average cost of sending $200 - 7%; Banks -11%) and unofficial channels dominate the flow. Remittances reduced poverty rates modestly in some countries and can act as insurance against disasters or economic downturns, allowing households to better smooth their consumption (Schaeffer, 2009; Castles, de Hass and Miller, 2014). The impact of remittances on education may be particularly important for economic development in the long run. Remittances also appear to finance entrepreneurial activity.

Aid

The ability of the Southern nations to finance themselves as well as the availability of increased financing from alternative sources has reduced the role of aid. While aid may still have some influence in the 34 remaining LICs, its importance in the MICs is negligible (Table 13.5).

The BWIs are not the only or the most important provider of official development assistance. There are now three major groups: The OECD-DAC donors, the Arab World and China and the other non-DAC donors (Figure 13.1).

The Arab donors concentrate their aid in Islamic countries – there are now 57 members of the Islamic Development Bank (IsDB). Patterned after the BWIs, these bilateral donors and the IsDB provide regional and country-specific aid in economic and social development emphasizing the achievement of the SDGs.

China's development assistance has generated much interest. It seems to be more transactional based on a neo-mercantilist quid-pro-quo. China has used aid to pursue commercial advantages including access to natural resources as well as foreign policy considerations such as favorable votes in the UN (Aiddata, 2020). Interestingly, China's preference for infrastructure projects have generated much support in the LICs and LMICs who view them as essential for economic

TABLE 13.5 Net ODA dependency 2017

Income groups/ Countries	Per capita aid ($)	Aid (% of GNI)	Aid (% of gross capital formation)	Aid (% of imports)
Low Income (LIC)	73	9.6	26.1 (2016)	27.1
LMIC	14	0.7	2.4	2.2
UMIC	11	0.1	0.2	0.3
Bangladesh (LMIC)	15	1.4	4.9	6.3
Colombia (HMIC)	23	0.3	1.3	1.2
Egypt (LMIC)	−1	0.0	−0.3	−0.1
Ghana (LMIC)	43	2.2	9.7	5.0

Source: World Development Indicators, 2019

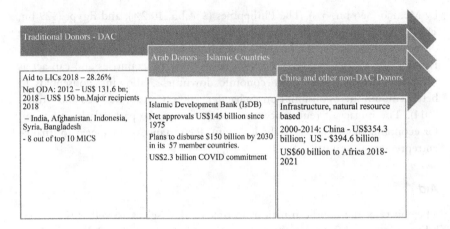

FIGURE 13.1 Aid patterns

Sources: World Bank, https://data.worldbank.org/indicator/NY.GDP.MKTP.KD.ZG?locations=TH
Islamic Development Bank (ISDB), https://www.isdb.org

Aiddata (2020), Tracking China, https://magazine.wm.edu/issue/2020-spring/tracking-china.php?
ref=aiddata

development (Ibid: 2020). This is different from the Washington consensus conditionalities that the North demands in exchange for aid.

Chinese assistance to LICs and MICs can be viewed as an instrument for hegemonic domination. There seems little likelihood that China will accede to current neoliberal international norms and standards. However, the new hegemonic constructs being created by China and other Southern donors can also lead to unequal power relationships between them and the recipients.

New development financing entities

The nature of development partnerships is also changing. The BWIs currently engage in both LICs and MICs. However, instruments used to solve "low income problems" are not appropriate for addressing "middle income issues". The latter will need new and innovative mechanisms. This is generally not the case now (Rahman and Barayni, 2018).

The MICs have different development interests oriented towards sustainable economic development and greater global integration. An increasing number of these countries are ready for and want loans, guarantees and equity financing – rather than grant funding – to boost infrastructure and finance economic growth including the growth of the private sector (as the principal revenue earner and employment generator). Impact investing, social businesses, specific mechanisms for promoting private investment are illustrative of attempts to adjust to these new realities. One option for the North and South governments and institutions is to become the facilitators of dialogues and interactions such as South-South and triangular Cooperation rather than being the primary funder (Rahman and Barayni, 2018).

Virtually all donors – multilateral and bilateral – now have Development Finance Institutions (DFIs), whose sole purpose is to promote the growth of the private sector. The latest one is the US International Development Finance Corporation (DFC) who "partners with the private sector to finance solutions to the most critical challenges facing the developing world today" (US International Development Finance Corporation: 2020). The World Bank equivalent is the International Finance Corporation (IFC) and the IsDB has the Islamic Corporation for the Development of the Private Sector (ICD). The DFIs typically provide equity and debt financing with some providing risk insurance as well.

China has embarked on a grand hegemonical scheme that weaves in development financing. President Xi has promoted a vision of a resurgent China (Economy, 2014). At the same time, a slowing down of the economic growth rate, the emergence of domestic excess capacity as well as strained trade relations with the United States have led to a search for alternative markets (Cuervo-Cazurra and Genc, 2008).

The BRI, sometimes referred to as the New Silk Road, is the major external response to this dynamic. Launched in 2013, it is one of the most ambitious infrastructure projects in history. The BRI stretches from East Asia to Europe, creating a vast network of railways, energy pipelines, highways, and streamlined border crossings, both westward through the former Soviet republic and southward, to Pakistan, India, and the rest of Southeast Asia. Such a network could also expand the international use of Chinese currency (Chatzky and McBride, 2020). More than sixty countries – with two-thirds of the world's population – have signed on or indicated an interest in doing so. China has already spent an estimated $200 billion with the largest project being the $68 billion China-Pakistan Economic Corridor connecting China to Pakistan's Gwadar Port on the Arabian Sea. China's overall expenses over the life of the BRI could reach US$1.3 trillion by 2027 (Chatzky and McBride, 2020). The BRI has its critics – the projects are loan based and could cause a debt problem and many investments are tied to the use of Chinese firms (Chatzky and McBride, 2020).

The AIIB can be viewed as a financing instrument for the BRI. Established in 2015, it aims to boost infrastructure connectivity in Asia and beyond and has 102 members (AIIB, 2020). Its governance structure appears to be relatively egalitarian with one Governor appointed by each member country. To date, AIIB has approved 83 projects in 23 countries in energy (25%), Financial institutions (20%), transport (18%), urban infrastructure (17%) and water (12%) with approved financing of US$19.14 billion and committed financing of US$11.25 billion (AIIB, 2020). The Institution has pledged to be "lean, clean, and green", but has financed mega fossil fuel and hydro projects.

The BRICs Bank represent another component of China's grand vision in creating an alternative international framework. The sixth BRICS Summit in 2014 established the New Development Bank (NDB) essentially as a counterpoint to the World Bank. The purpose of the NDB is to strengthen cooperation among BRICS and supplement the efforts of multilateral and regional financial institutions for global development (NDB, 2020). NDB's initial authorized capital was US$100 billion and subscribed capital was US$50 billion equally shared among the founding members. Its headquarters is in Shanghai. The NDB has a regional center in South Africa and should open another in São Paulo by the end of this year. It has established partnerships with other development banks, such as IADB, AIIB, and the World Bank group. The NDB has approved 42 projects in member countries with a total value over US$12 billion (thethirdpole.net, 2019). An early assessment of development financing from the BRIC countries to LICS found that BRICs lend more to LICs with weaker institutions and relatively rich in resources (Mwase: 2011).

Conclusion: The future of development financing and Covid-19

Southern countries are increasingly taking ownership of their development processes and reducing their dependency on international development assistance. The increasing weight of the alternative donor groups and institutions like the NDB and AIIB have reduced space for their Northern counterparts. Finally, the anti-globalization and isolationist postures of many Northern countries have resulted in reduced interest and hence funding for Southern development.

In response, the BWIs have been trying to remake themselves. There has been a tendency for these institutions to concentrate on global public goods like climate change, health pandemics and humanitarian assistance that appeal to Northern constituencies. The one noticeable exception are the DFIs promoting private sector growth. Independent bilateral aid agencies are disappearing and becoming parts of foreign ministries, the latest casualty being the Department of International Development (DFID) in the UK. While in principle, such mergers are suited to forming whole of government partnerships with the MICs, they also disadvantage the LICs in that they become lower-priority partners.

Globally, it is unclear what construct will emerge out of the current multi-polar chaos. One important conundrum is how to balance competing interests with the need for international codes of conduct. At a recent conference on the future of development financing, Mark Malloch-Brown suggested that "We have arguably the most disappointing set of governments globally that we have had in a very long time. That isn't just a coincidence of elections – it's a world that's deeply frustrated by the failure of the last generation of governments to address growing problems of inequality, exclusion, and marginalization of people" (Donback, 2019).

The BRICS will play important regional roles. However, this will not be conducive to setting any international behavioral norms. This vacuum is now being filled by global market mechanisms where the private sector particularly the MNCs are essentially free to chart their own course. The question is whether this is a good solution from a social welfare perspective in that private capital markets are not designed to provide the type of patient capital needed to address the development challenges.

The impact of the Covid-19 pandemic

As of July 2, 2020, there were about 11 million corona-positive cases in the world causing more than 500,000 deaths – among the developing regions, Africa had 318,432 cases with 6,340 deaths, Americas had about 2.5 million cases (Brazil 1,402,041) and about 125,000 deaths, with Southeast Asia accounting for 883,735 cases and 22,769 deaths; India had 604,641 cases and 34,788 deaths (WHO, 2020). There were no remedies for this virus at the time.

The Covid-19 lockdowns have created tremendous economic shockwaves. The World Bank baseline forecast envisioned a 5.2 percent contraction in global GDP in 2020 – the deepest global recession in decades, despite extraordinary efforts of governments to counter the downturn with fiscal and monetary policy support (World bank, 2020a). Advanced economies were projected to shrink 7 percent and the emerging market and developing economies by 2.5 percent – their weakest showing in at least sixty years. These downturns could reverse years of development progress causing tens of millions of people to fall back into extreme poverty. Another important feature was the historic collapse in oil demand and oil prices.

The fiscal and monetary stimulus' to offset the impacts of the pandemic have triggered historic government deficits – the US budget deficit increased by US$1.88 trillion (Washington Post, 2020); the UK budget deficit could increase to 300 billion pounds or around 15% of annual economic output (Miliken, 2020); Germany plans to increase new borrowing by 62.5 billion euros to 218.5 billion euros this year (Reuters, 2020); and the Canadian annual federal deficit is likely to be Can$260 billion (CBC, 2020).

This means that the availability of international development financing will reduce significantly. The first reason is the immediate deficit financing requirements of both Northern and Southern nations. The longer-term impact when the pandemic is over, will occur as the governments start paying back their debts to reduce deficits. Most Southern nations will have to fend for themselves for the foreseeable future and many of them may lose much of the economic and social gains achieved in the last decades.

A secondary impact may be more hegemonical. Historically, responses to global disasters have been led by the US and its northern allies. However, "one characteristic of the current crisis has been a marked lack of US leadership. The United States has not rallied the world in a collective effort to confront either the virus or its economic effects" (Haass, 2020). In contrast, despite the fact that the virus originated in China, it has stepped into the leadership void to provide material assistance and share information about the pandemic and China's experience battling the disease (Campbell and Doshi, 2020). This represents a further corroboration of the accelerated attenuation of the unipolar US led neoliberal global order and a signal of the emerging multi-polar world.

Notes

1 The North is defined as the US and Western European countries.
2 The South is defined as the countries in Africa, the Americas and Asia.
3 The ITO charter was ratified in 1944. But, the charter was not ratified by the US Senate. A watered-down version of the ITO with lesser norm setting and regulatory powers, the General Agreement on Tariff and Trade (GATT) was adopted in its place, subsequently replaced by the World Trade Organization (WTO) set up in 1995 as a result of the Uruguay Round of the Trade negotiations.
4 Alesina and Dollar (2000) find significant differences between donors. Nordic donors respond more to poverty type incentives (income levels, good institutions, and openness); France to former colonies tied by political alliances, the US are vastly influenced by its interest in the middle east.
5 This is based on the World Bank income classification method based on per capita income in US$in 2016 (Atlas Method). In this classification, Low-income countries are those with per capital income of US$1045 or less; Low middle-income countries (MICs) per capita incomes between US$1046-4125; High MICs are between $4126-12746; and High-income countries are those with per capita income US$12747 or more.

References

AIIB (2020). www.aiib.org https://www.aiib.org/en/index.html

Alesina, A. and Dollar, D. (2000). "Who gives foreign aid to whom and why?" *Journal of Economic Growth*, 5(1), pp. 33–64.

Arita, S. (2013). "Do emerging multinational enterprises possess South-South FDI advantages? *International Journal of Emerging Markets*, 8(4), pp. 329–53.

Bretton Woods Project (2019), What are the main criticisms of the World Bank and the IMF, https://www.brettonwoodsproject.org/2019/06/what-are-the-main-criticisms-of-the-world-bank-and-the-imf/

Burnside, C. and Dollar, D. (2000). Aid, policies and growth. *American Economic Review*, 90(4), pp. 847–868.

CBC (2020), Federal deficit likely now at $260 billion due to COVID-19, PBO says, https://www.cbc.ca/news/politics/federal-deficit-pbo-prediction-1.5585960

Campbell, K.M. and Doshi, R. (2020), The Corona Virus Could Reshape Global Order, Foreign Affairs, https://www.foreignaffairs.com/articles/china/2020-03-18/coronavirus-could-reshape-global order?

Castles, S., de Hass, H and Miller, M. J. (2014), *The Age of Migration: International Population Movements in the Modern World*, Fifth Edition, Palgrave Macmillan

Chatzky, A. and McBride, J. (2020), China's Massive Belt and Road initiative, Council on Foreign Relations, https://www.cfr.org/backgrounder/chinas-massive-belt-and-road-initiative

Cuervo-Cazurra, A. and Genc, M. (2008). Transforming disadvantages into advantages: Developing-Country MNEs in the least developed countries. *Journal of International Business Studies*, 39(6), pp. 957–79.

Dollar, D. (2015), The AIIB and the "one Belt, one Road". Brookings, https://www.brookings.edu/opinions/the-aiib-and-the-one-belt-one-road/

Donback, N. (2019), The Future of Development Finance: 5 event takeaways, DEVEX, https://www.devex.com/news/the-future-of-development-finance-5-event-takeaways-95936

Easterly, W. (2003). Can foreign aid buy growth? *Journal of Economic Perspectives*, 17(3), pp 23–48.

The Economist (2018), South-to South Investment is rising sharply, https://www.economist.com/finance-and-economics/2018/02/08/south-to-south-investment-is-rising-sharply

Economy, E, (2014), China's Imperial President: XI Jinping Tightens His Grip, Foreign Affairs, https://www.foreignaffairs.com/articles/china/2014-10-20/chinas-imperial-president

Frank, A.G. (1966), The development of underdevelopment, monthly review, Volume 18, *1966 and Reprinted in Latin America: Underdevelopment or Revolution*, (Monthly Review Press, 1969)

Global Policy forum (2009), Voting Share at the IMF and the World Bank 2009, https://www.globalpolicy.org/component/content/article/104-tables-and-charts/46584-voting-share-at-the-imf-and-the-world-bank-2009.html

Haass, R. (2020), The pandemic will Accelerate History Rather Than Reshape It, Foreign Affairs, https://www.foreignaffairs.com/articles/united-states/2020-04-07/pandemic-will-accelerate-history-rather-reshape-it?

Hooghe, L. and Marks, G. (2005). Calculation, Community and cues: Public opinion on European integration, *European Union Politics December*, Volume 6, pp. 419–443, https://doi.org/10.1177/1465116505057816

Lind, J. and Press, D.G. (2020), American Power in an Age of Constraints, Foreign Affairs, https://www.foreignaffairs.com/articles/china/2020-02-10/reality-check

Milanovic, B. (2019), The Clash of Capitalisms – The Real Fight for the Global Economy's Future, Foreign Affairs, https://www.foreignaffairs.com/articles/united-states/2019-12-10/clash-capitalisms

Miliken, (2020), Explainer: How will Britain pay for coronavirus borrowing?, Reuters https://www.reuters.com/article/us-health-coronavirus-britain-borrowing/explainer-how-will-britain-pay-for-coronavirus-borrowing-

Mwase, N. (2011), Determinants of Development Financing Flows from Brazil, Russia, India and China to Low Income Countries, IMF Working Paper, Washington

NDB (2020) https://www.ndb.int/

Pickford, S. (2019). Renew the Bretton Woods System, Chatham House, https://www.chathamhouse.org/expert/comment/renew-bretton-woods-system

Rahman, S.S. and Barayni, S. (2018). Beyond binaries: Constructing new development partnerships with middle-income countries, *Canadian Journal of Development Studies*, Volume 39, Issue 2, pp. 252–269, https://www.tandfonline.com/eprint/iv3B3tMBGEDGVYfIxqGi/full[Mismatch]

Reuters (2020), Germany's debt plans create budget deficit of 7.25% this year – sources, https://business.financialpost.com/pmn/business-pmn/germanys-debt-plans-create-budget-deficit-of-7-25-this-year-sources

Ritzer, G. (2010), *Neo-Liberalism, Globalization: A Basic Text, Chapter 5*, pp. 109–135, Wiley-Blackwell, United Kingdom

Schaeffer, R.K. (2009), *Understanding Globalization, the Social Consequences of Political, Economic, and Environmental Change*. Lanham, Maryland, Rowman & Littlefield, 2009, Chapter 5, pp. 101–146

SDG Pulse (2020), Robust and Predictable Sources of Financing for Sustainable Development, https://sdgpulse.unctad.org/investment/

Subramanian, A. and Kessler, M. (2013). The hyperglobalization of trade and its future. *Ssrn Electron. J*, 27, pp. 319–341.

Thethirdpole.net (2019). What happened to BRICs Bank? https://www.thethirdpole.net/2019/11/18/new-development-bank/

UNCTAD (United Nations Conference on Trade and Development) (2016). Key Statistics and Trends in International Trade, New York and Geneva.

UNDP (2015), Financing development through better domestic resource mobilization, Our Perspectives, https://www.undp.org/content/undp/en/home/blog/2015/12/22/Financing-development-through-better-domestic-resource-mobilization.html

WHO (2020), WHO Coronavirus Disease Dashboard, https://covid19.who.int/?gclid

Washington Post (2020), U.S. budget deficit widened to $1.88 trillion over eight months amid spending blitz to combat coronavirus downturn, https://www.washingtonpost.com/business/2020/06/10/budget-deficit-treasury-coronavirus/

Weisbrot, M. and Johnston, J. (2016), Voting Share Reform at the IMF: Will it make a Difference, CEPR Center for Economic and Policy Research, https://cepr.net/images/stories/reports/IMF-voting-shares-2016-04.pdf

World Bank (2018), Piecing Together the Poverty Puzzle, World Bank, https://openknowledge.worldbank.org/bitstream/handle/10986/30418/9781464813306.pdf

World Bank (2018a), Global Investment Competitiveness Report 2017/2018: Foreign Investor Perspectives and Policy Implications, World bank, Washington

World Bank (2020), History- World Bank Group, www.worldbank.org

World Bank (2020a), The Global Economic Outlook During the COVID-19 Pandemic: A Changed World, https://www.worldbank.org/en/news/feature/2020/06/08/the-global-economic-outlook-during-the-covid-19-pandemic-a-changed-world

World Data Lab (2019), World Poverty Clock, https://worldpoverty.io/

Zaman, H. and Rahman, S. (2020). SDG tracking through SSC for better SDG spending, in: *The Dynamics of Economic Space: Better Spending for Localizing Sustainable Development Goals.* Fayyaz Baqir, Nipa Banerjee and Sunny Yaya (eds.). Routledge: London and New York.

14

TURNING THE TIDE ON CANADA, THE EMPIRE

Genuine reconciliation, pluriversality, and indigeneity

Carolyn Laude

"When you speak to most senior government officials, bureaucrats and captains of industry, they have been raised to believe Indigenous people have no rights. And they've had a history of 150 years of court decisions that have supported that very principle that Indigenous people have no rights and that their rights are non-enforceable in Canadian law. Therefore, when they see Indigenous people protesting against the taking of their land or demanding more recognition of their traditional territorial rights, they see that as an unreasonable position, as a position that is going to go nowhere and is just stopping good ideas and good development and is preventing them from making money".

> *Senator Murray Sinclair, speaking with journalist Tanya Talaga about the Wet'suwet'en conflict over pipeline development, February 29, 2020*

In the aftermath of the *Tsilhqot'in Nation* decision, there was elation within certain Indigenous communities at the possibility of "genuine reconciliation" (CBC News, 2014) between the Canadian state and Indigenous peoples. In this case, the Court reconciled access and benefit to land and resources over other considerations. However, the Court also conceded that "Aboriginal title"[1] is not absolute and can be justifiably limited in the event of a broader public goal such as general and other forms of economic development (Tsilhqot'in Nation 2014, para. 81 – 82; 84; Delgamuukw 1997, para. 165). This suggests a hollow conception of reconciliation in which the state's sovereignty supersedes Aboriginal title and the compelling and substantial public interest (ibid, para. 88) supplants Indigenous jurisdiction, laws, sovereignty and rights. The power of Aboriginal title, as a liberal right, therefore, seems to lie in what Radha D'Souza (2018) claims is its "power of promise" (p. 51). If what D'Souza says is true, then Canada's entire approach to reconciliation through the "recognition of rights, respect, co-operation, and partnership" (Government of Canada 2015) with Indigenous peoples may hold nothing more than empty promises.

Moreover, the Federal Government's advancement of Indigenous-state relations raises the pressing issue of how to reconcile contra-distinctive "worldviews"[2] or "lifeworlds"[3] which focus on Indigenous land, rights and sovereignty. While the "politics of recognition"[4] advances reconciliation through a plan of economic development, rights-oriented legal strategies, treaty-making and other constructive arrangements in the national and international realms (Government of Canada 2015; Coulthard 2014; Flanagan 2006), Indigenous peoples' claim inherent rights, relationality to land/place (Lindberg 2012, p. 91) and self-determination as the basis for a renewed relationship. A tension therefore arises between a "settler-colonial liberal-capitalist democracy"[5] ideology of reconciliation and that of Indigenous relationality to land/place wherein inherent rights flow from the Creator and not government laws (Lindberg 2012; Cardinal and Hildebrand 2000, p.11). This calls attention to dominant conceptions of land and rights needing to reconcile capitalist economic development with Indigenous relationality to land, rights, sovereignty and legal traditions. This is what constitutes reconciliation in Canada. Nonetheless, we must ask ourselves what the decolonial reconciliation possibilities are given Canada's imperialistic tendencies. Thus, how can decolonial views of reconciliation make possible a transnational pluriverse in Canada? And, what must Canada do to look out to the world stage from an Indigenous worldview?

To address these challenges, I will touch upon the present state of Indigeneity, reconciliation, and Indigenous rights in Canada. Moreover, I will offer a sketch of the potential of pluriversality to achieve genuine reconciliation. Exploration of reconciliation from this lens positions "ways of co-existing"[6] differently to re-conceptualize the pluriverse in Canada. It signals the importance of seeking decolonial forms of reconciliation. To comprehend the hollow proposition of settler-colonial reconciliation, I argue that Indigenous inherent rights and sovereignty are considered inferior given Canada's imperialistic tendency of making Indigenous rights and land important targets of capitalistic forces. This flags the significance of Indigeneity and how divergent understandings of land, jurisdiction and rights challenge Indigenous-state relations.

State of indigeneity, reconciliation and rights issues

I approach the subject of Indigeneity from a First Nation-centric lens, preferring to distinguish between "Eurocentrism"[7] and Indigenous epistemological-ontological modes. The debates surrounding Indigeneity have strong political and legal under-currents. As a result, Indigeneity in Canada is defined in terms of rights-holders and/or as established under the *Indian Act 1985,* essentializing Indigeneity into narrowly defined identity categories. Some Indigenous community members agree with the state's definition; however, others are confused and polarized by what Indigeneity truly means. Regardless, settler-colonialism uses race to control and/or make Indigeneity visible, including how it approaches reconciliation.

The bureaucratic and political buzz on reconciliation has produced a multitude of reactions. Public servants grapple with the meaning of reconciliation and how to put it into practice. Political messaging and actions, on the other hand, have provoked a range of First Nation reactions. Additionally, there is much disparity amongst Federal and Indigenous governments on how best to reconcile land and rights issues. Some First Nations agree with "...a renewed legal and political relationship of rights recognition" (Coulthard 2014, p. 3) in which resources and rights are redistributed through constructive arrangements. This view of reconciliation assimilates Indigenous polities into Canadian society and merely reproduces configurations of colonial power instead of "peaceful co-existence grounded in the ideal of mutuality" (Coulthard 2013, p. 2). Alternatively, some First Nations maintain the Indigenous lifeworld is the only possible framework and they do not engage with state institutional constructs. Nevertheless, others believe reconciliation is achievable through a hybrid model of the first two approaches. As a result, federal efforts to secure Indigenous buy-in of government policies and approaches are often contested.

In 2018, the resistance of Idle No More activists, First Nations and the Defenders of Land and Truth Network displayed a growing displeasure with a proposal to embed Indigenous rights into a rights-recognition framework. First Nations perceived government as "continuing its old ways", raising suspicions and fostering distrust (Brake 2019). Indigenous peoples felt their inherent rights, self-determination and jurisdictions were being threatened. Chief Trevor Mercredi stated: "The government has taken away the authority of the people, the true rights holders in Canada...They're domesticating everything we own..." (Brake 2019). Indigenous leaders and community members are now less tolerant of government's colonial shape-shifting tactics. The failure to secure First Nation support demonstrates resistance to Canada's dominion over the Indigenous lifeworld. Instead, First Nations are desirous of nation-to-nation relationships in which they are equal partners.

Reconciliation and transnational pluriversality

Despite efforts to address moral, political and legal issues resulting from wrongdoing and violence against Indigenous peoples, there is a question about whether Canada can reconcile deeply ingrained colonial difference. In earlier work, I examined reconciliation discourses and found that it can be understood, as follows: "genuine"[8], "stand-alone"[9], "transitional"[10] and "assimilative"[11]. The findings suggest reconciliation is a contested term. On the one hand, Canada's normative understanding of reconciliation includes legal and political mechanisms and on the other, de-colonial reconciliation focuses on "sub-altern"[12] epistemologies, cosmologies, and lived colonial experiences. The analysis demonstrated that genuine reconciliation held the most promise for finding a pluriverse path-forward.

Since genuine reconciliation embraces "pluriversality"[13], this implies Western epistemologies and cosmologies and claims of modernity are not the dominant narrative and solution (Grosfoguel 2016; Mignolo 2013). This form of reconciliation serves to: enact inter-cultural, legal, political, and epistemological change; reconstruct the structure-culture narrative; engage in relational responsibility and dialogue; balance power; value other worldviews; and, stabilize the culture-ecology-economic dichotomy.[14] Acceptance of diverse worldviews is critical to realizing genuine reconciliation. To move beyond the settler-colonial paradigm of reconciliation, all governments must think, be and act differently with respect to Indigeneity.

This is only made possible if government chooses to adopt a practice of genuine reconciliation wherein lifeworld co-existence embraces the pluriverse of Indigenous cosmologies and epistemologies and an "Nth-eyed seeing"[15] approach. As a reconciliation approach, Nth-eyed seeing makes space for diverse worldviews to jointly produce solutions aimed at improving human and social conditions. The concept is grounded in a philosophy of liberation and decolonization, pushing settler-colonial epistemological, philosophical, and structural boundaries.

Wet'suwet'en Hereditary Chief Na'moks' reflections on the importance of respecting the Wet'suwet'en lifeworld in pipeline development illustrates what it means to push boundaries in relation to achieving genuine reconciliation:

> … They want to talk about reconciliation? Well, I don't know if I ever want to use that word again… What the hell did we do except be Wet'suwet'en? Not once did we break our law. They said the people arrested were being civil disobedient, well they were being obedient to our law (Morin 2019).

According to Borrows (2016), civil (dis)obedience can be interpreted in two ways: obedience to "Indigenous peoples' law or the state's own unenforced or unrealized standards" (p.53). The first signals obedience to Wet'suwet'en law and respect for the nation's authority on Wet'suwet'en lands. Moreover, (dis)obedience suggests the prospect of a "third space of sovereignty" (Bruyneel 2007) wherein the truism of a settler-colonial lifeworld is opposed. Chief Na'moks infers the Indigenous lifeworld has a role to play in settler-colonial decision-making processes. As proposed, genuine reconciliation accepts that the Indigenous lifeworld has a legitimate role in Canada's political and legal system.

As a paradigm shift, genuine reconciliation situates Indigenous cosmologies and epistemologies and the "coloniality of the marginalized"[16] at the forefront. The role of Indigenous lifeworlds is strengthened instead of conceding to the dominant geo-politics of knowledge mode in which "there can be no others" (Mignolo 2002, p. 59). According to Fanon (1967), this shift is one of "damné from the very perspective of the damné" (cited in Mignolo 2005, p. 395).

The damné experience, location, worldview and lifeworld thus shapes and informs genuine reconciliation.

Indigeneity, Canada, and the world stage

I cite Duncan Ivison's work as an example of what not do when recognizing Indigenous worldviews and rights in a plural context. His proposal fuses normative understandings of Indigenous rights within a system of "complex mutual co-existence"[17] to maintain cultural identity. I disagree with this proposition for several reasons. First, political efforts to embed one lifeworld into a dissimilar lifeworld will never result in deep structural reform. As a political construct, Federalism is grounded in racial privilege. Therefore, transnational and global subordination of Indigenous rights, laws, jurisdiction and sovereignty will continue regardless of the state's well-intentioned efforts. Evidence also suggests that a rights-recognition model will continuously infringe upon Indigenous socio-cultural interests due to their status as "negative rights"[18]. Under a settler-colonial political ideology, the natural environment and Indigenous socio-cultural interests have few, if any, rights.

Second, global capitalism dominates the Indigenous lifeworld and inherent rights. First Nations have expressed deep concern that international trade agreements erode their constitutional and treaty rights concerning land and environmental and cultural measures (Clogg and Sage 2012). *Hupacasath First Nation v. Canada* (2015) argued that foreign corporations had more power than communities over resource management and that investor rights could infringe upon Indigenous rights. The argument highlighted the *Canada-China Foreign Investment Promotion and Protection Agreement's* (FIPA) ability to impact negatively on asserted inherent title and rights (Patterson 2017). The court disagreed and concluded as follows:

> ... Canada did not owe Hupacasath a duty to consult because the alleged potential adverse impacts of FIPA on its interests were "non-appreciable" and "speculative", and Hupacasath had failed to establish any causal link between FIPA and its alleged impacts (*Hupacasath First Nation v. Canada* 2015, para. 8).

This ruling is curious, but not surprising. It contrasts the Supreme Court of Canada view, which encourages "an expansive view of section 35 and the promise of a balanced relationship between the Crown and Indigenous peoples" (Justice Canada 2017). In other words, the Crown must act honourably and participate in processes of negotiation with Indigenous peoples (Haida Nation 2004, para 25).[19] This is not reflexive of the Hupacasath experience. However, Corntassel and Woons (2018) believe that global power sharing is made possible by "reinvigorating treaties with the natural world, (re)establishing alliances between Indigenous peoples and/or Indigenous advocacy in diplomatic activities within global forums" (p. 5). This suggests genuine reconciliation can arise when the Indigenous lifeworld and pluriversality are part of global power-sharing arrangements; however, this is only possible to attain if Canada rejects its sovereign-to-subject position.

Moreover, the use of non-discrimination and expropriation clauses in FIPA style agreements position investor rights above Indigenous rights to land. Manuel and Schabus (2005) confirm that "Indigenous peoples are in direct competition with multi-national corporations for control over their lands" (p.229). Additionally, the Government of Canada requires restitution when Indigenous peoples' actions negatively affect a foreign company's future profit-making. The Tsawwassen First Nation Final Agreement includes a clause to address lost revenue situations, as follows:

> 31. Where ... a Tsawwassen Law or other exercise of power by Tsawwassen First Nation causes Canada to be unable to perform an International Legal Obligation, ... Tsawwassen First Nation will remedy the Tsawwassen Law or other exercise of power to the extent necessary to enable Canada to perform the International Legal Obligation (Clogg and Gage 2012, p.3).

Domestication of Indigenous rights questions the legitimacy of having constitutionally protected section 35(1) rights in Canada. As Manuel and Schabus (2005) argue, these agreements privilege multi-national wealth accumulation and highlight Indigenous peoples' inability to "secure the implementation of their ancestral land rights" (p. 229). Canada's globalization efforts therefore reproduce colonial behaviours that erode Indigenous rights and relationality to land. As a colonial tool, they reinforce a power imbalance that upholds the sovereign-to-subject ideology, nationally and globally. According to Corntassel and Woons (2018), present-day inter-state relations constitute nothing more than "violence, broken treaties and other unjust assertions of power over Indigenous peoples and their lands" (p. 1). This situation reinforces the presence of "systemic or structural racism"[20] in Canadian laws, treaties, policies, operational practices and institutions.

Canada, the empire

According to Todd Gordon (2010a), Canada uses capitalism as a pathway to acquire power and influence over Indigenous peoples within its nation-state boundaries (p.17). The Canadian state exploits Indigenous land and labour as part of an imperialist strategy to ensure wealth creation (Gordon 2009, p. 48). Imperialist relations of "power and domination" (Gordon 2010a, p. 16) are most revealing in Canada's approach to resource development and Indigenous rights from 2012 to 2015. At that time, the Federal Government aggressively promoted responsible resource development (RRDI) and created an economic action plan to address 600 resource extraction projects (Government of Canada 2012a). Peck and Tickell (2002) describe this occurrence as the rolling-out of neoliberalism policy in which the market is a vehicle to ensure "job readiness" and mobilize marginalized groups for labour (p. 392). Responsible resource development was foundational to Canada's survival, commandeering Indigenous land and labour for wealth generation.

Indigenous scholar Glen Coulthard (2014) claims the practice of dispossession or primitive accumulation drives the "productive character of colonial power... and coercive authority..." (p. 152). In support, Todd Gordon (2010b) maintains that Canadian imperialist efforts rob Indigenous peoples of their natural resources, wealth, land and rights through treaty-making, exploitation or theft as a function of accumulation by dispossession. For Gordon (2010b), primitive accumulation is the result of accelerated capitalist exploitation, which is detrimental to Indigenous peoples. This suggests that legal and political interventions often serve to further constrain Indigenous sovereignty, laws, rights and jurisdiction. In other words, Canadian laws, policies and practices control and limit Indigenous peoples' inherent rights and their relationality to land.

Although jobs and resource development are essential tools for lifting Indigenous peoples out of poverty, Coulthard (2013) observes that: "colonial powers will only recognize the collective rights and identities of Indigenous peoples insofar as this recognition does not obstruct the imperatives of state and capital" (p. 8). This signals the impossibility of achieving "sovereignty at par" between Indigenous and non-Indigenous political orders. Thus, Canada's agenda as empire discounts the Indigenous lifeworld and its inherent rights, resulting in negative consequences for Indigenous peoples that do not achieve genuine reconciliation.

Conclusion

Genuine reconciliation is the only viable alternative to settler-colonial reconciliation and imperialistic tendencies. By recognizing the existence of pluriversality, Indigenous lifeworlds are respected. Emphasis is placed on co-existence and emancipation as tools for sharing space with Indigenous worldviews and inherent rights in national and global arrangements. If this is true, then genuine reconciliation centers on co-creating knowledge and co-developing a new moral, political and legal pluriversal order. By reconciling many worldviews/lifeworlds, this offers the possibility of transforming settler-colonial institutions to balance power, bridge the racial divide, and emphasize Indigenous humanity, rights and socio-political and legal agency instead of wealth creation. Therefore, genuine reconciliation acts as a tool to decolonize and improve power-sharing and collaborative problem-solving locally, nationally and globally.

Notes

1 Aboriginal title is a sui generis communal right that is site, fact and group specific. It is grounded in historical evidence that Indigenous peoples had control over the land prior to European contact. An assumption of Crown sovereignty confers underlying title to the Crown for all land in Canada (Delgamuukw 1997, para. 1017).

2 For Leroy Little Bear (2000), worldviews arise from a culture or a society's shared philosophy, values and social customs. This speaks directly to people's interaction and inter-dependency with the land, the animals and the people surrounding them (p. 77).

3 Aaron Mills (2016) speaks of lifeworld as "a set of ontological, cosmological, and epistemological understandings which situate us in creation and thus which allow us to orient ourselves in all our relationships in a good way" (p. 852). He contends that lifeworlds begin with First Nation creation stories (p.855). They also include Indigenous political and legal ordering and rights.

4 Coulthard (2014) interprets the politics of recognition as "recognition-based models...reconciling Indigenous assertions of nationhood with settler-state sovereignty via the accommodation of Indigenous identity claims in...a renewed legal and political relationship..." (p.3).

5 I consider settler-colonialism and liberal democracy as inseparable. Therefore, I refer to this embeddedness as settler-colonial liberal-capitalist democracy. This concept focuses on political economy, which is tied to capitalism and labour. Liberalism takes the form of a centralized nation-state that upholds personal autonomy, reasonableness and a liberal society of "like-minded individuals" (Hueglin 1994, p. 10). Citizens experience "freedom and equality of opportunity" in the form of nation-state control over their living conditions (ibid, pp. 9 - 10). This means citizens are required to accept a certain level of compromise and subservience to the nation-state (ibid, pp. 9 - 10). Conversely, some individuals and/or groups experience liberalism not only as "oppressive standardization and conformity", but also as a form of "subordination to, and exploitation by, centrally dominant elites" (ibid, p.10).

6 For Alfred (2005) co-existence is "...morally grounded defiance and non-violent agitation...to generate within the settler society a reason and incentive to negotiate... in the interest of achieving a respectful coexistence" (p.27). This theory of liberation parallels that of the de-colonial project.

7 Pokhrel (2011) describes Eurocentrism in terms of how non-Western histories and cultures are understood from a Western perspective. It relies on the assumption that European cultural values were superior and universally applied to diverse societies despite their social, cultural and historical differences (p.9; Quijano 2000; Dussel 1998).

8 Although the decolonial scholarship does not espouse a theory of reconciliation, I interpret pluriversality as a potential pathway for co-existence. Acceptance of diverse worldviews instead of the dominant worldview is critical to realizing genuine reconciliation.

9 Stand-alone reconciliation refers to an Indigenous "oppositional, place-based existence" that redeploys Indigenous culture and values (Coulthard 2014; Simpson 2011). This view rebalances community relationships and decolonizes nation-to-nation political and economic relationships. This form of reconciliation restores Indigenous people's humanity, dignity and cultural/spiritual strength, the ecology and provides enough land to realize self-determination (Alfred 2005). Lastly, it reflects Indigenous nationalism, failing to account for the global matrix of power in a colonial-capitalist world-system.

10 Transitional reconciliation implies Indigenous reliance upon makeshift spaces of sovereignty to resolve their instability inside and outside of the settler-colonial system (Henderson 2013; Bruyneel 2007, p.14). Reconciliation is realized in these third spaces of resistance where Eurocentric theories of culture and categories of identity are critiqued, and economic and cultural values are redefined and redistributed (Henderson 2013; Bruyneel 2007). Use of settler-colonial systems and processes is encouraged despite Indigenous claims to autonomy. This type of reconciliation is distinct from anti- and de-colonial reconciliation, which disapprove of embedding Indigenous lifeworlds within the dominant lifeworld given its racist and sexist foundation.

11 Assimilative reconciliation privileges settler-colonial values and institutions over those of the Indigenous lifeworld, favouring economic relations over ecological and social relations to support an economic logic of profit-making.

12 Sub-alterns are persons/groups that are socially, politically and geographically marginalized by the nation-state's hegemonic power structure in imperial and modern/colonial global world contexts (Mignolo 2005, p.387). Indigenous peoples in Canada are considered sub-alterns.

13 Pluriversality consists of inter-connected cosmologies arising "through and by the colonial matrix of power" (Mignolo 2013, p.2). Coloniality means control of the economy, authority, gender and sexuality, and subjectivity and knowledge (Quijano 2000).

14 Racial, political and social hierarchical ordering of Indigenous peoples under the *Indian Act* is a form of coloniality. Indigenous voices are necessary to balance power and to effect structural reform to address the deep inequality and oppression experienced by Indigenous peoples. Genuine reconciliation is concerned with protecting nature, equal distribution of wealth and labour and promoting resistance to assimilative policies.

15 Nth-eyed seeing is designed for the Canadian setting; however, it can be adapted to other local, national and global situations. The concept brings pluriversal worldviews into conversation with each other. Nth-eyed seeing is rooted in *Etuaptmumk or Two-Eyed Seeing*, bringing the strength of Indigenous and Western approaches together. It is useful when thinking about new possibilities and seeking long-lasting solutions that can benefit all parties (Bartlett et al. 2007, para. 1).

16 This refers to the way in which human beings are controlled under a classification system of race, gender, class, and labour and property (Mignolo 2002; Quijano 2000). Race and racism are its organizing principles. Coloniality plays a role in Canada's colonial process through the universal imposition of European beliefs and values upon Indigenous peoples including economic coloniality from which capitalism arises.

17 Ivison (2010) draws from the Two-Row Wampum teaching to explain what complex mutual co-existence means, focusing on mutual recognition, respect, and peaceful co-existence (cited in Duthu 2013, p. 56).

18 Graham (2011) claims negative rights can be violated, purchased and gifted at will (p.185). Compensation and/or other incentives address the infringement or violation of negative rights.

19 The Honour of the Crown governs the Crown's conduct in its dealings with Indigenous peoples from "...assertion of sovereignty to the resolution of claims and the implementation of treaties..." (Haida 2004, para. 17).

20 Hughes (2020) argues that systemic or structural racism is "inherent in institutions" (p.1). Canada and other countries have built social, economic and political systems based on dominant norms and values. The policies, practices, cultural representations and laws of a country are the backbone of these systems. Even if unintentional, relationships are based on dominance and subordination of one lifeworld over another.

References

Alfred, T. (2005). *Wasáse: Indigenous Pathways of Action and Freedom*. Peterborough: UTP Broadview Press.

Bartlett, C., Marshall, M. and Marshall, A. (2007). Integrative Science: Enabling Concepts within a Journey Guided by Trees Holding Hands and Two-Eyed Seeing [online], unpublished manuscript part of Two-eyed Knowledge Sharing Series. Cape Breton University, March 2007. Available from: http://www.integrativescience.ca/uploads/articles/2007-Bartlett-Marshall-Integrative-Science-Two-Eyed-Seeing-Aboriginal-co-learning-trees-holding-hands.pdf [access on 15 October 2019].

Borrows, J. (2016). *Freedom & Indigenous Constitutionalism*. Toronto: University of Toronto Press.

Brake, J. (2019). Resistance mounts at AFN forum in Edmonton as Canada continues with planned policy overhaul. *National News* [online], 1 May 2019. Available from: https://aptnnews.ca/2019/05/01/resistance-mounts-at-afn-forum-in-edmonton-as-canada-continues-with-planned-policy-overhaul/ [accessed 12 October 2019].

Bruyneel, K. (2007). *The Third Space of Sovereignty: The Postcolonial Politics of U.S. – Indigenous Relations*. Minneapolis: University of Minnesota Press.

Cardinal, H. and Hildebrand. (2000). *Treaty Elders of Saskatchewan: Our Dream Is That Our Peoples Will One Day Be Clearly Recognized as Nation* [online]. Calgary: University of Calgary Press, pp. 11 - 30. Available from: Scholars Portal Books: Canadian Electronic Library [accessed 30 October 2019].

CBC News. (2014). Tsilhqot'in First Nation Granted B. C. Title Claim in Supreme Court Ruling. *CBC News* [online], 26 June 2014. Available from: http://www.cbc.ca/news/politics/tsilhqot-in-first-nation-granted-b-c-title-claim-in-supreme-court-ruling-1.2688332 [accessed 28 June 2014].

Clogg, J. and Sage, A. (2012). Is Canada-China Investment Treaty (FIPA) an attack on Aboriginal rights? West Coast Environmental Law Blog [online], 30 October 2012. Available from: https://www.wcel.org/blog/canada-china-investment-treaty-fipa-attack-aboriginal-rights [accessed 12 October 2019].

Corntassel, J. and Woons, M. (2018). *Expansion Pack: Indigenous Perspectives on International Relations Theory* [online]. Canada: International Relations Publishing, pp. 1 – 5. Available from: https://www.e-ir.info/2018/01/23/indigenous-perspectives-on-international-relations-theory/ [accessed 08 September 2020].

Coulthard, G. (2013). Indigenous peoples and the politics of recognition [online]. *New Socialist Webzine*, pp. 1 – 8.

Coulthard, G. (2014). *Red Skin, White Masks: Rejecting the Colonial Politics of Recognition*. Minnesota: University of Minnesota Press.

Delgamuukw v. British Columbia 1997. S.C.J. No. 108, 3 S.C.R. 1010 (S.C.C.).

Department of Justice. (2017). *Principles Respecting the Government of Canada's Relationship With Indigenous Peoples* [online]. Ottawa: Government of Canada. Available from: http://www.justice.gc.ca/eng/csj-sjc/principles-principes.html [accessed 20 August 2018].

D'Souza, R. (2018). *What's Wrong With Rights?: Social Movements, Law and Liberal Imaginations*. England: Pluto Press.

Dussel, E. (1998). Beyond eurocentrism: The world-system and The limits of modernity. In: Jameson, F., and Miyoshi, M., eds. *The Cultures of Globalization*. Durham: Duke University Press, pp. 3 - 31.

Duthu, B. (2013). *Shadow Nations: Tribal Sovereignty and the Limits of Legal Pluralism*. Oxford: Oxford University Press.

Flanagan, T. (2006). *First Nations? Second Thoughts*. Montreal: McGill-Queen's University Press.

General, L. The Redman's Appeal for Justice. Letter to the Honourable Sir James Eric Drummond, 1923. Held in League of Nations, document no. 11/30035/28075. Palais des Nations, Geneva.

Gordon, T. (2009). Canada, empire and indigenous people in the americas. *Journal of the Society for Socialist Studies*, pp. 47–75.

Gordon, T. (2010a). Canada's imperialist project. *Briarpatch*, 29(3), pp. 16–19. Available from: Gale Academic OneFile [accessed 30 September 2019].

Gordon, T. (2010b). *Imperialist Canada*. Winnipeg: Arbeiter Ring Publishing.

Government of Canada. (2012a). Responsible Resource Development [online]. Available from: http://actionplan.gc.ca/en/page/r2d-dr2/frequently-asked-questions-responsible-resource-development#q4)> [accessed 24 August 2019].

Government of Canada. (2015). *Minister of Crown-Indigenous Relations and Northern Affairs Canada Mandate Letter* [online]. Ottawa: Government of Canada. Available from: https://pm.gc.ca/en/mandate-letters/minister-crown-indigenous-relations-and-northern-affairs-mandate-letter [accessed 10 October 2019].

Graham, N. (2011). *Lawscape: Property, Environment, and Law*. New York: Routledge.

Grosfoguel, R. (2016). Decolonizing the Academy I: Decoloniality, Transmodernity and the World System Seminar [unpublished]. Decolonizing Post-Colonial Studies and the Paradigms of Political Economy Series. University of Edinburgh, 24 - 26 February.

Haida Nation v. British Columbia (Minister of Forests) 2004. SCC 73.

Henderson, J. S. (2013). Incomprehensible Canada. In: Henderson, J., and Wakeham, P. eds, *Critical Perspectives on the Culture of Redress* [online]. Toronto: University of Toronto Press, pp. 115 – 126. Available from: Scholars Portal Canadian University Presses [accessed 25 July 2019].

Hueglin, T. O. (1994). Exploring Concepts of Treaty Federalism: A Comparative Perspective [online]. Available from: http://data2.archives.ca/rcap/pdf/rcap-81.pdf [accessed 20 February 2018].

Hughes, P. (2020). From Discrimination to Systemic Racism: Understanding Societal Construction. *SLAW Canada's Online Legal Magazine*, 30 June [online]. Available from: http://www.slaw.ca/2020/06/30/from-discrimination-to-systemic-racism-understanding-societal-construction/ [accessed 3 September 2020].

Hupacasath First Nation v. Canada (Attorney General) 2015. FCA 4.

Indian Act (R.S.C., 1985, c. I-5).

Justice Canada. (2017). Principles respecting the Government of Canada's relationship with Indigenous peoples. Ottawa: Government of Canada. Available from: http://www.justice.gc.ca/eng/csj-sjc/principles-principes.html [accessed 12 August 2018].

Little Bear, L. (2000). *Chapter 5: Jagged worldviews colliding*. In: Battiste, M. ed, *Reclaiming Indigenous Voice and Vision*. Vancouver: University of British Columbia, pp. 77–85.

Manuel, A. and Schabus, N. (2005). Indigenous peoples at the margin of the global economy: A violation of International human rights and International trade law. *Chapman Law Review* [online], 8, pp. 222 – 251. Available from: https://www.chapman.edu/law/_files/publications/CLR-8-arthur-manuel-and-nicole-schabus.pdf [accessed 23 September 2020].

Mignolo, W. (2002). The geopolitics of knowledge and the colonial difference. *The South Atlantic Quarterly* [online], 101 (1), pp. 57 – 96. Available from: http://www.unice.fr/crookall-cours/iup_geopoli/docs/Geopolitics.pdf [accessed 04 October 2019].

Mignolo, W. (2005). On subalterns and other agencies. *Postcolonial Studies*, 8(4), pp. 381–407. Available from: Scholars Portal Journals: Taylor & Francis Current [accessed 10 September 2019].

Mignolo, W. (2013). On Pluriversality *De-colonial Thoughts Op-eds* [online], pp. 1 – 4. Available from: http://waltermignolo.com/on-pluriversality/ [accessed 17 October 2019).

Lindberg, T. (2012). The doctrine of Discovery in Canada. In: Miller, R., Ruru, J., Behrendt, L. and Lindberg, T. ed, *Discovering Indigenous Lands: The Doctrine of Discovery in the English Colonies*. Oxford: Oxford University Press, pp. 89 - 124.

Mills, A. (2016). The lifeworlds of law: On revitalizing indigenous legal orders today. *McGill Law Journal*, 61(4), pp. 847–884. Available from: HeinOnline Law Journal Library [accessed on 15 October 2019].

Morin, B. (2019). Wet'suwet'en hereditary chief: 'Reconciliation is not at the barrel of a gun. National Observer [online], 16 April. Available from: https://www.nationalobserver.com/2019/04/16/news/wetsuweten-hereditary-chief-reconciliation-not-barrel-gun. [accessed 17 July 2019].

Patterson, B. (2017). Indigenous rights at risk as Canada-China Free Trade Agreement talks begin February 20. *The Council of Canadians Blog* [online], 17 February 2017. Available from: https://canadians.org/blog/indigenous-rights-risk-canada-china-free-trade-agreement-talks-begin-february-20 [accessed 18 September 2020].

Peck, J. and Tickell, A. (2002). Neoliberalizing space. *Antipode*, 34(3), pp. 380–404. Available from: Wiley Online Library [accessed 20 December 2015].

Pokhrel, A. K. (2011). Eurocentrism. In Chatterjee, D. K. ed, *Encyclopedia of Global Justice* [online]. Dordrecht: Springer. Available from: https://link.springer.com/referenceworkentry/10.1007/978-1-4020-9160-5_25 [accessed 10 October 2019].

Quijano, A. (2000). Coloniality of power, eurocentrism, and latin America. *Nepantla: Views from South*, 1(3), pp. 533–580. Available from: Project Muse Premium Collection [accessed 19 April 2018].

Simpson, L. (2011). *Dancing on Our Turtle's Back: Stories of Nishnaabeg Re-Creation, Resurgence and a New Emergence*. Winnipeg: Arbiter Ring Publishing.

Talaga, T. 2020. Reconciliation isn't dead. It never truly existed. *The Globe and Mail*, 29 February [online]. Available from: https://www.theglobeandmail.com/opinion/article-reconciliation-isnt-dead-it-never-truly-existed/ [accessed 10 August 2020].

The Guardian. (2020). Canada: protests go mainstream as support for Wet'suwet'en pipeline fight widens. The Guardian [online], 14 February 2020. Available from: https://www.theguardian.com/world/2020/feb/14/wetsuweten-coastal-gaslink-pipeline-allies [accessed 14 February 2020].

Tsilhqot'in Nation v. British Columbia, SCC 44 2014. 2 S.C.R. 256.

15

CIVIL SOCIETY AND THE FAULT LINES OF GLOBAL DEMOCRACY

Fayyaz Baqir

Context: the social landscape of democratic societies

Democracy has emerged as the dominant form of government in the past half century. The number of people living under democratic rule increased from 12 percent of the global population in 1900 to 63 percent in 2000. However, democracy is more than just putting in place an elected government. Democratic order must also encompass economic and social democracy. It builds on democratic values and a broad array of citizens' voluntary associations. It also calls for erasing the boundaries between national and international "democracy" meaning that the idea of one person one vote should be part of a global practice. It cannot be reduced to majority rule and the arbitrary power of the majority. Democracy also means equality and accountability outside the seat of government, anchored in equitable relationships across the boundaries of class, caste, gender, race, and faith, in civic life, at the workplace and in neighborhoods, increasing the range of choices people have in their daily lives (Sen 1999). The normative legitimacy of democracy is based on the premise that anyone who is affected by a decision should also be a part of making that decision. Diversity creates new challenges for democratic politics because a majority decision cannot provide an answer to the deprivations caused by the structural foundations of discrimination. The perpetual existence of such discrimination has created disillusionment with democracies and has led to calls for democratizing democracies. In transitional democracies persistence of these exclusionist structures has been accompanied by electoral violence, dynasty politics and a culture of patronage (Swain, Amer and Öjendal 2009).

The events which unfolded after the death of handcuffed Black citizen George Floyd on May 25, 2020 in Minneapolis, Minnesota at the hands of a white police officer who pressed his knee on Floyd's neck for 8 minutes such that he could no longer breathe brought the numerous fault lines of the global democratic order under the public gaze. The wave of protests starting in Minneapolis spread like a wildfire across the globe. The protestors demanded not only the end to police brutality but the legacy of slavery and racism, economic discrimination, and social injustice. It was ironic to note that the global protests found resonance everywhere except Palestine. The reason was that such brutalities were a daily affair in Palestine and world conscience has been silent on this "new normal." This shows that a deep division exists not only between oppressive structures and vulnerable groups but between vulnerable people themselves. These protests led to dismantling of statues of Gandhi, Belgian King Leopold who massacred 1.5 million Africans in Congo, American President Andrew Jackson, and many other slave masters. These protests clearly emphasized the existence of deep structural divides in leading democracies.

Another earth-shaking event preceding the 2020 Black Lives Matter (BLM) protests was the spread of COVID 19 across the global. This event brought to public attention the interdependence of all the sections of the global community on the one hand, and the segregated healthcare systems across the globe on the other. Many leading democracies like the USA, UK and India were caught completely unprepared to deal with this disaster and miserably failed to protect the health, security, and safety of their citizens especially the weakest, poor and most vulnerable. Some of the so called "undemocratic" regimes like China and Vietnam handled the crisis much more effectively and with a much lower loss of life. The pandemic also highlighted the fissures in the global governance system, depicted for example by the withdrawal of US support for World Health Organization (WHO). Despite the collective nature of the threat many people in the political and business class saw the pandemic as an opportunity to make profits by calling for commercial solutions. The pandemic also revealed the fractured relationship of the global community with nature. In-depth review of the pandemic revealed that continued human encroachment of natural habitat in the name of development has created the dynamic of recurrent emergence and spread of such viruses. This brought to public attention the issue of climate change in a forceful way and revealed the well-entrenched conflict between scientific truth and commercial interests in the neoliberal order.

COVID 19 and BLM invite us to go beyond the unilateral concept of democracy as election of representatives to the pluralist concept of creating a world where many worlds can exist in harmony (Escobar 2019). Pluralism implies a system for making citizens a part of power structure to go beyond majority vote and identity based power structures, and find ways to create a world where many worlds can exist simultaneously. This concept of democracy follows the principal of a globalization- based on pluralism not the hegemony of rentier capital. It affirms globalization based on fraternity of sovereign local economies as part of the tapestry of the global economy (Escobar 2019). A true deep-rooted democratic system can be established if we can move from the concept of representative to participatory democracy (UN 2004). It is also important to note here that true democracy's function is to make peace (Iber 2018). But this peace cannot last without justice. Social justice has emerged as a key issue in traditional democracies, democracies in transition and in countries without multiparty electoral frameworks. In the case of the latter, China agreed to the continuation of the justice system and civil liberties in Hong Kong for a period of 50 years after the end of British rule under its proclamation "One Country Two Systems" but started tracking back by introducing legislation for the extradition of dissenters to the mainland in 2019 leading to mass protests (The New York Times 2019). In Xinjiang province, the Chinese government started separating school children from their families, faith and language in the name of "Centralised Care" and "Thought Education" by sending them to giant boarding schools and held adults in detention camps for "vocational training" to fight extremism (BBC 2019). On the other hand, the question of social justice gains critical significance in traditional democracies because social injustice is in conflict with the norms upheld in the name of free market and democratic politics.

It seems that what electoral democracies have given people with one hand has been taken away by the other even under glorious democratic regimes. Democracy in communities, democracy in social life, democracy for minorities and the vulnerable, democracy at the workplace have all been questioned in Western democracies. Workplaces in the West are seen by many as private governments which function as despotic entities. For example, non-unionized workers have no recourse in the case of abuse, even if it is illegal like sex abuse or wage theft, other than quitting their jobs. Wage theft by corporate employers exceeds the sum total of all other thefts in the U.S. The majority of workers are forced to sign agreements for mandatory arbitration which deprives them of the right to be heard by a neutral judge. This control also extends to workers private lives. A Coke worker, for example, was fired for drinking Pepsi at lunch. Thousands of slaughterhouse workers cannot use the bathroom during work hours. They are told to wear diapers to work. Such abuses are legitimized in the name of "freedom of contract," meaning that workers can leave jobs if they do not like the workplace. While workers in the gig economy are called independent contractors, they are denied the benefits offered to workers doing similar jobs and are regulated as much as the regular employees in a firm (Anderson 2017). Corporations and elites have undermined government's capacity to respond to citizens' concerns by finding legal ways to defy human rights at the workplace (Reich 2009; Sachs 2011). The key questions we need to ask are: is the game is rigged and how can we create a level playing field?

The social fault lines of the democratic order

As noted by Thomas Picketty, the "Social State" went into retreat during the neoliberal era and its role as a handyman of the rich got entrenched in popular consciousness. The only factors that threaten this role are mass violence and progressive taxation (Sachs 2011; Mann 2020). This compels us to understand why the ideals of social justice, peace and human welfare cannot be achieved through electoral politics in democratic societies. To understand this problem, we need to look at the social fault lines in democratic societies and the limitations of electoral politics in building bridges across social divides.

Racism

Racism is a major source of discrimination in democratic societies. This was evident during the coronavirus pandemic when medical professionals in the US pointed out that race provided a major explanation of a person's likelihood of dying from the disease. Death rates in counties with large Black populations were found to be 10 times higher than the average country level death (Begley 2020). Police brutality is another dimension of racism. It has led to the loss of public trust and public approval of police forces, especially due to use of police as a military force in Black neighborhoods. Hate culture among white border patrol officers who formed a Facebook group to exchange racist comments is another example of how this trust has been bruised (Thompson 2019). While the economic cost of slavery has been computed by many scholars, whether African Americans will receive any reparations is anybody's guess. But black people have been economically hurt in other ways as well. It has been estimated that due to discriminatory federal policies in the US black farmers lost 12 million acres of land during the 20th century. These policies affected 1 million farmers constituting 98 percent of Black agricultural landowners. Many of these land theft practices were legal and based on flawed federal policies (Newkirk and Vann 2019).

Black Lives Matter ignited similar movements in India in the aftermath of police brutality against women, youth and minorities protesting against discriminatory laws (Kamdar 2020). Another way of embedding discrimination in electoral politics is asking citizenship questions in the American census to access data to redraw electoral districts along racial lines, giving over representation to white votes and under representation to minorities (Berman 2019). Racism has plagued not only American politics, but the political culture of the biggest democracy in the world, India, and the "only democracy in the Middle East," Israel. Harassment of minorities in the name of citizenship proof in India and construction of the wall in the West Bank to deny Palestinian farmers access to their land are two such examples. Background checks are used against Black workers in US to discriminate against them in hiring. Many states and municipalities have banned these practices because they affect access to jobs, housing, and other public services for 100 million Americans. It is difficult to stop these discriminatory practices in the temporary job market. Background checks tilt the scale against Black Americans because the ratio of African Americans with a criminal record is 33 percent against national average of 8 percent. Workers' complaints of wage theft and horrific work conditions against leading employers like Walmart or Schneider Logistics have been denied due to a maze of subcontracts between the workers and employers (Burns 2019).

Low-caste communities in India, though formally included in the democratic polity are structurally excluded from power structures. They are not even extended the dignity of being recognized as human beings. Many of the scavengers in the lower castes have to manually collect human feces from latrines, and numerous deaths are reported when they drown cleaning the deep septic tanks or gutters. There is no compensation in the case of death. They are not allowed to start small businesses. The legal and judicial system does not offer them any way out of this horror (Barton 2019). They are not spared racist vitriol even if they succeed in rising above their status to academic and administrative positions (Nigam 2019). "Liberal Racism" or micro aggression has also been reported in the form of everyday slights and indignities in British universities. It also takes the form of inaccessibility to "white" rules or networks. Non-white academics feel that that they have to be exceptional to be accepted as ordinary (Sian 2019).

Babasaheb Ambedkar, leader of India's untouchable caste, demanded a separate electorate for untouchables like Muslims to ensure their share in power during the British rule in India. Indian National Congress leader Mohandas Karamchand Gandhi threatened to fast till death if Ambedkar did not withdraw his demand. He was later inducted in the government as law minister after independence by Jawaher Lal Nehru, but he died a bitter and disillusioned man. To pull his community out of the vicious caste system he asked them to convert to Buddhism. One million of his followers converted. He accused Gandhi of hypocrisy and double dealing. No wonder protesters dug out similar charges against him during the Black Lives Matter protests in 2020 and attacked his statutes in Washington and London. Earlier his statue was removed from University of Ghana on charge of racist behavior against Africans (Ambedkar 2011; Safi 2018). While Gandhi has a global image of a prophet of non–violence he came under attack for his racist politics in South Africa whereby he pleaded for caste like segregation at the Durban post office, telegraph office and jails and demanded military training for Indians so that they could fight along with fellow white colonialists against AmaZulu tribesmen (Kambon 2018).

Another example of brazen racism is the "democratic" government of Israel's displacement of Palestinians from their land, their confinement as refugees under military occupation in Gaza and massacre of people who carry out peaceful protests against these brutalities. Israel projects the image of being the only democratic regime in the Middle East while they deny the right of peaceful protest to Palestinians under their occupation. The Great March of Return by Palestinians ethnically cleansed from the occupied territories that started in the spring of 2017 was met with sniper fire, in which children and unarmed Palestinians were mercilessly killed. Since the start of second Intifada in 2000 around 12,000 Palestinian children have been detained and interrogated by the Israeli army. It is one of the most consistent features of Israeli occupation since 1967 (Scheer 2019; Baroud 2019). The Israeli government also constructed a wall to separate Palestinians in the West Bank from their farms and pastoral lands. While, an International Court of Justice (ICJ) advisory issued on July 9, 2004 declared construction of the West Bank wall by Israel to be illegal and called Israel to dismantle the wall and pay reparations for any damages caused, the global conscience seems to be silent on the issue (Hong 2019).

Gender

Gender discrimination is another common practice in contemporary democratic societies. Gender discrimination is part of every form of discrimination. It becomes conspicuous in democratic societies that uphold the principle of one person one vote. A case worth mentioning is of the American Women's Soccer team who won the World Cup final in year, eclipsed the men's team as a lucrative franchise and went on to launch a lawsuit against the United States Soccer Federation to address the issue of equal pay. In protest the team's captain Rapinoe decided to receive honorable recognition not from the President of the United States (POTUS) but Congresswoman Alexandra-Ocasio-Cortez (Figueroa 2019).

Indigenous communities

The rights of Indigenous people, local communities and the environment have been violated even under left wing regimes in Greece, and parts of South America in pursuit of extractivist, growth led economies. This phenomenon has erased the difference between democratic and dictatorial regimes (Kothari 2019). It has called for bringing ethics back to politics. The principles of a community based ethical democracy call for participation, service, humility, submission to the community's will, use of reason not force, construction not destruction, and power to the people not politicians (Leyva-Solano 2019). More specifically it means reciprocity, appreciation of interdependence, and respect for unity in diversity and a preference for horizontal over hierarchical systems. In political terms it means confederal municipalism (Tokar 2019). Under this perspective the process of change draws our attention to the nature of interdependence in political discourse and this insight helps us to change social relations without taking power. It also ends the duality between society and nature (Gudynas 2019). It calls for an end to a culture of development based on performativity, desire as the defining principle of freedom as practiced in consumer democracies and upholds direct democracy. Governor of California Gavin Newsom likened the treatment of Indigenous people to genocide and apologised to Indigenous peoples for the crimes committed against them. This was in complete contrast to the statement in 1851 of the first governor of California, Peter Burnet, who said that the war against native people would continue until they were extinct (Oxford 2019).

The violation of the commons and silencing dissent on official environmental policies of modern nation states has been another denial of the democratic rights of people demanding mitigation of climate change, food sovereignty and protection of nature and life on earth. It is worth noting here that between 2002 and 2018 1,738 environmentalists were killed across 50 countries. Many of the deaths were related to the conflicts around to fossil fuels, water, timber, and natural resources (Butt and Menton 2019).

The idea of direct democracy must be accepted critically, however. As Zografos has pointed out "The Appenzell-Innerrhoden a Swiss canton that is celebrated as an example of direct democracy conceded voting rights to women only in 1991when forced by the Swiss Federal Supreme Court" (Zografos: 2019, 12). Interestingly Direct Democracy has gradually evolved as part of the urban planning practice in China: "What we have witnessed is an interactive process in which residents navigate their agency by both adapting existing space and creating new space, without seriously challenging the imbalance of power relation with authority" (Zhang et al. 2018, 1556).

Mass surveillance of citizens

Mass surveillance of citizens is another issue. Edward Snowden's moral rebellion is the case in point. Snowden gave up his future at the age of 29 because his conscience did not allow him to be part of the mass surveillance system that he helped create for the National Security Agency (NSA) and got disillusioned when no one among his colleagues paid attention to his ethical concerns about invading the privacy of millions of individuals in the name of national security (Chan 2019).

Capture and control of institutional powers for subduing dissenting voices has also emerged as a major political strategy of Indian political elites. A whistle blowing police officer was given life imprisonment in Gujrat and many journalists, lawyers and human rights activists have been thrown in jail (Philipose 2019). As pointed out by Major Danny Sjursen, the US government, with bipartisan support, is targeting whistleblowers and has escalated a war on freedom of the press (Sjursen 2019). A Canadian Civil Liberties Group the B.C. Civil Liberties Association (BCCLA) also accessed through a court order evidence of Canadian Intelligence Agencies spying on Enbridge Northern Gateway Pipeline protestors (Freeze and Bailey 2019).

State bodies on the other hand want to operate secretly. Chelsea Manning refused to testify before a grand jury investigating WikiLeaks founder Julian Assange as a protest against the practice of maintaining secrecy on issues of public interest. (Maxwell and Chakravarti 2019). Israeli Intelligence Agencies have also been found to have engaged in spying against the Boycott, Divestment and Sanctions movement in Israel and in covert activities to recruit social media warriors. Israel even declined the visa applications of US Congresswomen Ilhan Omar and Rashida Tlaib to visit the West Bank and East Jerusalem. Public Security Minister Gilad Erdan proposed sanctions against activists who oppose government policy toward Palestinians. Use of citizen's social credit is now emerging as a new form of governance (Szeto 2019, Goldberg 2019).

Mass media, corporate finance and "Manufacturing consent"

Corporate power has commodified political discourse and meaningful civic engagement has been replaced by political consumerism and the gaze of the mainstream media. A cascade of media campaigns continues before and after the election only to be interrupted on election day. Political parties have turned into money laundering operations for the super-rich and social media duties are farmed out to media outlets. Electoral politics has turned into the normalization of spectacular media events and meaningful engagement has been converted into a consumption experience.

Education and inequality

The education system that played the role of great equalizer and created a vibrant middle class in the US does not function that way anymore. All the benefits of economic growth have been captured by the big corporations and their share-holders in the past few decades. While after tax profits of corporations doubled from 5 percent to 10 percent of the GDP after 1970, the share of wages declined by 8 percent of GDP. As a result, $2 trillion USD was shifted from the mid-dle class to the super rich. In the case of poor sections, the situation is worse. Children of low-income families have to frequently change school due to poor housing, have little help with homework, have chronic health problems and live in chaotic and unsafe neighborhoods (Hanauer 2019). Student debt is another burden which has been carried by baby boomers, Gen-Xers, Millennials, and Gen-Z students. Repayment of these debts is playing havoc with their lives. In the US, homeownership by Millennials has declined by 20 percent, a significant number have postponed marrying and having children. The situation is similar in other democracies. The nuclear family, the building block of "free societies," seems to be falling apart (Abrams and Cody 2019). At the same time wealthy parents have been bribing coaches and helping their children to cheat in exams to get them admission into college. Fifty people were charged by federal prosecutors in the US for paying $25 million to college-prep professionals to bribe coaches and administrators (Williams 2019).

Global economic ethics

The contemporary global economy is following a set of rules that have threatened the social fabric of the global village and put the survival of the planet at risk. Democratic politics in these times needs to rebuild an edifice of ethical economics to survive these challenges. Democratic societies need to create an economy where everyone equally shares in economic success; not only the shareholders of transnational corporations and their management (Szreter, Cooper and Szreter 2019). That is why US Supreme Court Justice Louis D. Brandeis very eloquently said, "We can have democracy in this country, or we can have great wealth concentrated in the hands of the few, but we can't have both" (Giroux 2019, n.p.). The ethics of contemporary economic rules are so flawed that victims of the 9/11 attack have not received compensation for their treatment even two decades after the attack. While the US government has spent trillions of dollars to capture the mastermind of the attack in Afghanistan it has not felt obliged to take care of the victims at home. Even comedian Jon Stewart, former host of The Daily Show, reprimanded the US Congress for its shameful inaction on the issue in a statement in 2019 (Siddiqui 2019). Rising inequality and stagnating incomes led to the dramatic upsurge in civil unrest in a quarter of all countries in 2019 (Dehghan 2020).

The extent of economic inequality caused by the corporate practice of tax evasion can be gauged by the fact that 40 percent of multinational corporations evade taxes amounting to $200 billion USD a year that are transferred to tax havens (Wier 2020). The rise of the Gig economy is another dimension of rising inequality. Two impacts of the gig economy in journalism are worth noting: it has deprived hundreds of thousands of journalists of their jobs and fair wages and ushered in the era of fake news dues to the loss of editorial vetting of content as more and more jobs in the commercial sector have been outsourced. This absolves the employers of their responsibilities to workers to protect them against workplace injury and harassment (Vigo 2019). The American healthcare system is in a state of disaster and 50 million people go uninsured every year (Bruenig 2019).

Social divide, identity politics and the civil society

Deep divisions in democratic countries have given rise to social movements, identity politics and the rights-based discourse of civil society organizations. The rise of identity politics in the 1980s broadened the scope of resistance against economic and social inequalities but created divisions within social movements. Blaming and shaming became major tools of identity politics and weakened the rational discourse and solidarity of the global precariat. Rising above identity politics has emerged as a major issue for social movements today (Consolo 2019; Fantasia 2019). During the last 40 years civil society organizations became very active in dealing with the issue of social discrimination. Civil society engagement on issues of citizen rights has been very strong in Western democracies, especially the US. As noted by Alexis de Tocqueville, the distinguishing feature of American democracy is not voting rights but the universal presence of free voluntary associations (Meurer 2019). Civil society organizations have played a key role in addressing shortcomings in human development, dispensing economic justice, alleviating poverty, mitigating climate change, ensuring sustainable development, re-appropriating the commons, and establishing sovereign community economies in view of market and public policy failures.

Civil Society's secret to success is presented in various names: participatory development, rights-based discourse, trickle down and bottom up approaches, social entrepreneurship, the voluntary economy, social capital, micro initiatives, and pluralist practices to mention a few. It is asserted that civil society succeeds where the market and government fail. In my view civil society can play a critical role in building trust and bridging gaps between various interest groups by creating a shared knowledge space – a prominent form of social capital – carved out for strengthening interactions between different identity groups on the one hand and between these groups and political and business classes on the other hand. CSOs engage in very diverse activities in dealing with the issues of social discrimination and structural exclusion. This diversity is the source of civil society resilience. Adal Najam has depicted this diversity in a meaningful way. He has pointed out that the space between the state and civil society is defined by the ends and means they define in relation to their objectives. These ends can be in harmony or conflict and the means available to them have a critical impact on defining the nature of their interaction. Depending on the nature of their ends and means civil society and the state or corporate capital can engage in four different ways. They may cooperate, confront, complement, or co-opt each other. If they have similar goals and similar means it will lead to a cooperative relationship; in case of similar goals and dissimilar means they will complement each other's efforts; in the case of dissimilar goals and similar means one party might co-opt the other and in case of dissimilar goals and dissimilar means, CSOs claiming rights will lead to confrontation (Najam, 2000)

Civil society can help deepen the democratic order by following a pluralist discourse. This entails a perpetual struggle to overcome its own limitations. For example, while global NGOs claim to be working for just global governance, the relationship between Northern and Southern NGOs comes under question due to the unequal nature of the relationship between them. While UN agencies, the global development community and international NGOs have been talking about "thinking globally and acting locally" the divide between Northern and Southern NGOs and development assistance agencies has left little space for Southern NGOs to define their own agenda and receive financial assistance for working on their own priorities. There is also limited transparency in the sector. For example, the whereabouts of 30 percent of the funds received for humanitarian assistance in Syria remain unknown. 99.1 percent of NGOs in the Global South work as subcontractors of Northern NGOs which implies working on a predetermined agenda as well. Many donors insist that they are under pressure to channel funds through their own NGOs. It is interesting to note here that out of 3000 NGOs engaged in development of Sustainable Development Goals (SDGs) meant to be implemented mostly in the Global South, less than 100 or 3.3 percent are from the South. Northern governments have also been accused of hiding behind the tendering process of Request for Proposals (RFPs) to shape friendly economic and social policies (Okumu 2019).

Conclusion: Pluralism, the path of agency and transformation

Political Agency comes with transformative capacity (Baqir 2007; Patomaki 2020). The ineffectiveness of states and markets as agents for local development arises in dealing with the social reality of low-income communities and marginalized groups. Market and Government Agency fails to transform their living conditions, their human, social and financial assets and undermines their sovereignty for managing their own lives and habitats. This world can be changed not through new templates but new processes. Templates reproduce existing relationships. The process of accessing, creating, and documenting local knowledge empowers communities to deal with the inflexibility of templates and their incompatibility with the ground reality.

My contention is that "means do exist for the population to free themselves but the option has remained undeveloped" (Sharp 2011, p. 13) and that is the weakness of civil society and donor funded NGOs engaged in poverty alleviation, improvement of livelihoods, right based development and good governance. Realistic options according to Sharp include, one, political defiance i.e. withdrawal of popular and institutional cooperation. This entails knowledge and understanding of mutual dependence of the state and citizens. It also entails knowledge of the Achilles' heel of the powers that be: not being able to act without cooperation and support of the precariat. Two, it includes negotiation. For employing any of these options Sharp has identified four possible mechanisms, i) conversion of the power holders, ii) accommodation of mutual interests, iii) non-violent coercion and iv) disintegration of oppressive power (Sharp 2011). The most important concern raised by Sharp is about designing a *grand strategy*. In the situation of extreme fear and powerlessness he proposes extremely low risk and confidence building measures to start defying the system defiance. Strategists should choose the issues that receive wider recognition and provide minimal ground for rejection. They may begin defiance with selective resistance, by starting battles that they can most likely one, thereby building the confidence of communities in defiance and preparing them to fight bigger and more risky battles.

Non-violent resistance in the first place should enable communities and the state to move from the morality of fear to the morality of hope. The dynamics of this transformation have been described by Sharma in an account of an English Police Officer dealing with the protestors. The officer stated "I do not like your people, and do not care to assist them at all. But what am I to do? … I wish you took to violence like the English strikers, and then we would know how to dispose of you. But you will not injure even the enemy. You desire victory by self-suffering alone and never transgress your self-imposed limits of courtesy and chivalry. And that is what reduces us to sheer helplessness" (Sharma 2013, p. 71). The problems in communities are caused by market and government failure which is closely connected with moral deficit (Leitner 1982). This moral deficit takes the form of not sharing information, using discretionary powers, not tapping local knowledge, not dispensing duties as prescribed, using public office for private benefit, and giving preference to prescription over process. This moral deficit undermines agency and strengthens the perpetuation of existing social relationships. The panacea for socio-economic ills is not civil society but inclusiveness.

Examples of movements based on inclusion and trust include some very inspiring initiatives, especially those taken by women. These initiatives include: the Green Belt Movement in Kenya started by Wangari Maathai in which rural women planted trees as part of a soil and conservation effort to avert desertification of their land; the Akwesasne Mother's Milk, Project Mohawk established by women along the St. Lawrence River to monitor PCB toxicity while continuing to promote breastfeeding as a primary option for women and their babies; the Greening of Harlem initiated by Bernadette Cozart, a gardener and founder who organizes diverse community groups in Harlem to transform vacant garbage-strewn lots into food and flower gardens; Sister Rivers performance ritual in which Japanese women placed rice, seeds, and soil from Hiroshima and Nagasaki in pillowcases and then floated the artwork down the Kama River; the exposure of the Love Canal as a toxic waste site set off by Lois Gibb and her founding of the Citizens Clearinghouse for Hazardous Waste to share tactical skills with local environmental groups; and the Chipko movement where women stopped the felling of trees by hugging the trees, standing between the axe and the tree. The moral principle of pluralist social change is that you find truth by embracing the "other."

References

Abrams, Natalia and Cody Hounanian (2019) Life, deferred: student debt postpones key milestones for millions of Americans, Open Democracy 14 August 2019 https://www.opendemocracy.net/en/oureconomy/life-deferred-student-debt-postpones-key-milestones/

Ambedkar, B.R. speaks on M.K. Gandhi (BBC Radio) April 3, 2011 https://www.youtube.com/watch?v=_FNSQcEx02A

Baqir, Fayyaz, *UN Reforms and Civil Society Engagement*, UNRCO, Islamabad, 2007

Barton, Naomi (2019) *No Dignity, No Rights, But Filth Forever: Manual Scavengers in Photographs*, The Wire, 4 July 2019

Baroud, Ramzy (2019) *The War on Innocence: Palestinian Children in Israeli Military Court*, Counter Punch, August 9, 2019

BBC News 4 July 2019. China Muslims: Xinjiang schools used to separate children from families https://www.bbc.com/news/world-asia-china-48825090 1/19

Begley, Sharon (2020) To grasp who's dying of Covid-19, look to social factors like race – STAT. June 15, 2020 https://www.statnews.com/2020/06/15/whos-dying-of-covid19-look-to-social-factors-like-race/

Berman, Ari (2019) GOP Docs Prove Census Citizenship Question Is About Preserving White Political Power, Democracy Now 04 June 2019-Transcript Reader Support News

Bruenig, Matt (2019) *The US Health System Is a Nightmare Where 50 Million Go Uninsured Every Single Year,* Jacobin., 02 July 19

Burns, Rebecca (2019) *Black Workers Say Walmart's Background Checks Are Racially Discriminatory,* In These Times 05 May 19

Butt, Nathalie and Menton, Mary (2019) More than 1,700 activists have been killed this century defending the environment. *The Conversation,* August 5, 2019

Chan, DM (2019) *Snowden Memoir Set for Fall Launch Former National Security Agency Contractor Leaked Treasure Trove of Secret Documents in 2013,* Asia Times. August 1, 2019

Consolo, Olivier (2019) *Beyond the Choir: A Contribution to the GTI Forum: Farewell to the WSF?* GTI Forum, October 2019

Dehghan, Saeed Kamali *One in Four Countries Beset by Civil Strife as Global Unrest Soars | Global development | The Guardian,* 16 Jan 2020

Escobar, Arturo (2019) *The Global Doesn't Exist: A Contribution To The GTI Forum: Think Globally, Act Locally?* Great Transition Initiative, August 2019

Fantasia, Rick (2019) "In America, identity replaces solidarity" Le Monde diplomatique - English edition, August 2019

Figueroa, Meleiza (2019) OP-ED- ECONOMY & LABOR, Brilliant Defiance: Megan Rapinoe and AOC Are the Future, Truthout, published July 8, 2019, (https://truthout.org/authors/meleizafigueroa/),

Freeze, Colin, and Bailey, Ian (2019) CSIS kept records on environmental groups, interacted with oil industry officials, files show, *Globe and Mail* July 8, 2019

Goldberg, Emma (2019) *This Is How Israeli Democracy Ends* – The Forward, August 15, 2019

Giroux, Henry A. (2019) Let's Shut Down the Authoritarian Machine, Truthout, June 19, 2019 https://truthout.org/articles/lets-shut-down-the-authoritarian-machine/

Gudynas, Eduardo (2019) Value, growth, development: South American lessons for a new ecopolitics, *Capitalism Nature Socialism,* 30:2, 234-243, DOI: 10.1080/10455752.2017.1372502

Hanauer, Nick Better Schools Won't Fix America, The Atlantic, July 2019 issue Hong, Chung-Wha (2019) Resisting the Wall Industry, From Mexico to Palestine, Yes Magazine, July 10, 2019

Iber, Patrick (2018) *Worlds Apart: How Neoliberalism Shapes the Global Economy and Limits the Power of Democracies* The New Republic April 23, 2018

Kambon, Obadele (2018) *Ram Guha Is Wrong. Gandhi Went from a Racist Young Man to a Racist Middle-Aged Man,* The Print, 24 December 2018

Kamdar, Bansari (2020) *Addressing Race in India and Abroad: Colorism, Surveillance, and Reckoning With Police Impunity,* The Diplomat, June 16, 2020

Kothari, Ashish (2019) ed. *Dimensions of Democracy,* Kalpavriksh, Pune,

Leitner, G.W. *History of Indigenous Education in the Punjab Since Annexation and in 1882. Calcutta 1882.* Reprint Delhi: Amar Prakashan, 1982.

Leyva-Solano, Xochit in Zapatista Autonomy in Ashish Kothari (2019) ed. *Dimensions of Democracy,* Kalpavriksh, Pune,

Mann, Geoff (2020) The Inequality Engine-LRB vol. 42 No. 11 4 June 2020

Maxwell, Lida and Chakravarti, Sonali, (2019) Chelsea Manning Against the Grand Jury, Jacobin, June 10, 2019

Meurer, Michael (2019) Elections Are Treated as Political Consumerism and Democracy Is Suffering-Op-Ed Truthout, August 6, 2019

Najam, A. (2000). The four C's of government third sector-government relations. *Nonprofit Management and Leadership,* 10(4), pp. 375–396.

Newkirk, II and Vann, R. (2019) *"The Great Land Robbery: The Shameful Story of How 1 Million Black Families Have Been Ripped from Their Farms."* The Atlantic, September 2019

Nigam, Aditya (2019) Statement of academics on Rabindra Bharati University incident of harassment, Kaflia online 28 June 2019

Okumu, Paul (2019) *"How NGOs in Rich Countries Control Their Counterparts in Poor Countries and Why They Refuse to Resolve,"* Naked Capitalism, July 15, 2019

Oxford, Andrew (2019) *California Governor Calls Native American Treatment Genocide,* Associated Press, June 18, 2019

Patomaki, Heikki (2020) *Conjuring the Spirits of the Present,* Forum: Planetize the Movement! Great Transition Initiative, April 2020

Philipose, Pamela (2019) Backstory: Shrinking Spaces Need Expanding of Awareness 22 June 2019 https://thewire.in/media/backstory-shrinking-spaces-need-expanding-of-awareness 2/12

Reich, Robert B. (2009) How capitalism is killing democracy, *Foreign Policy,* October 12, 2009

Sachs, Jeffrey D. The Global Economy's Corporate Crime Wave Project Syndicate Apr 30, 2011 https://www.project-syndicate.org/commentary/the-global-economy-s-corporate-crime-wave?barrier=accesspaylog

Safi, Michael (2018) *Statue of "Racist' Gandhi Removed from University of Ghana,* The Guardian, 14 December 2018

Sian, Katy (2019) *Extent of Institutional Racism in British Universities Revealed through Hidden Stories,* The Conversation, June 27, 2019

Sen, Amartya (1999) *Development as Freedom,* Oxford University Press

Sharma, Arvind (2013) *Gandhi- A Spiritual Biography*; Yale University Press, New Haven 2013

Sharp, Gene: *From Dictatorship to Democracy,* Serpent's Tail, London 2011

Scheer, Robert (2019) We Could Solve the Israel-Palestine Conflict Tomorrow, Truthdig, July 25, 2019

Siddiqui, Sabrina (2019) *Jon Stewart Demands Congress Act for 9/11 Responders: "They Did Their Jobs-Do Yours'* The Guardian. UK. 11 June 2019

Sjursen, Danny (2019) A Soldier's Defense of Chelsea Manning and Julian Assange, truthdig, 11 June 2019 https://www.truthdig.com/articles/a-soldiers-defense-of-chelsea-manning-and-julian-assange/

Swain, A, Amer, R and Öjendal, J. (Eds.). (2009). *The Democratization Project: Opportunities and Challenges.* London; New York; Delhi, Anthem Press. Retrieved July 2, 2020, from www.jstor.org/stable/j.ctt1gxp8mw

Szeto, Hannah (2019) *Mossad Involvement in Israel Anti-BDS Campaigns Exposed.* Middle East Monitor, June 12, 2019

Szreter, S, Cooper, H and Szreter, B (2019) *Incentivising an Ethical Economics: A Radical Plan to Force a Step Change in the Quality and Quantity of the UK'S Economic Growth,* IPPR Economics Prize. http://www.ippr.org/economics-prize/

The New York Times, *Carrie Lam, Hong Kong Leader,* Condemns Violence, July 2, 2019

Thompson, A. C. (2019) - *Border Patrol Refuses to Come Clean on Secret Facebook Group.* Truthdig July 11, 2019 https://www.truthdig.com/articles/border-patrol-refuses-to-come-clean-on-secret-facebook-group/

Tokar, Brian in Zapatista Autonomy in Ashish Kothari (2019) ed. *Dimensions of Democracy,* Kalpavriksh, Pune,

UN Report of the Panel of Eminent Persons on United Nations–Civil Society Relations entitled "We the peoples: civil society, the United Nations and Global Governance." New York, June 2004

Vigo, Julian (2019) *The Gig Economy and Outsourcing: A Dark Net of Near Slavery*, Truthdig, September 7, 2018

Wier, Ludvig (2020) Tax havens cost governments $200 billion a year. It's time to change the way global tax works, *World Economic Forum* 27 February 2020

Williams, Walter E. (2019) Higher education fraud not limited to admissions scams – *Atlanta Journal-* 12 May, 2019

Zhang, M., Wu, W. and Zhong, W. (2018). Agency and social construction of space under top-down planning: Resettled rural residents in China. *Urban Studies*, 55(7), 1541–1560. https://doi.org/10.1177/0042098017715409

Zografos, Christos- Direct Democracy in Ashish Kothari (2019) ed. *Dimensions of Democracy*, Kalpavriksh, Pune

INDEX

Note: *Italicized* and **bold** pages refer to figures and tables respectively.

Printed in the United States
by Baker & Taylor Publisher Services